GRAPHING CALCULATOR MANUAL
FOR WANER AND COSTENOBLE'S

FINITE MATHEMATICS

SECOND EDITION

Mark Stevenson
Oakland Community College

Florence Chambers
Oakland Community College

Australia • Canada • Mexico • Singapore • Spain • United Kingdom • United States

Sponsoring Editor: *Curt Hinrichs*
Assistant Editor: *Stephanie Schmidt*
Production Editor: *Scott Brearton*
Cover Coordinator: *Vernon Boes*
Marketing Team: *Samantha Cabaluna, Beth Kroenke*
Print Buyer: *Micky Lawler*
Cover Printing: *Webcom Limited*
Printing and Binding: *Webcom Limited*

For more information about this or any other Brooks/Cole products, contact:
BROOKS/COLE
511 Forest Lodge Road
Pacific Grove, CA 93950 USA
www.brookscole.com
1-800-423-0563 (Thomson Learning Academic Resource Center)

Printed in Canada

5   4   3   2   1

ISBN 0-534-37900-1

TI-83 Plus Graphing Calculator Supplement
by
Florence Chambers

TI-86 Graphing Calculator Supplement
by
Mark Stevenson

CHAPTER 1

LINEAR FUNCTIONS AND MODELS

TI-83 PLUS

```
LinReg
 y=ax+b
 a=-1.5
 b=3.5
```

1.1 Functions from the Numerical and Algebraic Viewpoints - TI-83 Plus

In this section you will see several methods for evaluating a function on the TI-83 Plus. The first method requires that you enter the function as if you were going to graph it, and the second method uses the TABLE of values generated by the graph of the function. In addition, this section shows you how to test a value in an inequality.

Before we begin with the work in this section, let's look at some important information about graphing functions on the TI-83 Plus.

SOME GRAPHING PRELIMINARIES

How to Graph Functions

You can graph a variety of types of functions. The TI-83-Plus allows you to input ten (10) functions in the Y= menu. It would be difficult to look at all ten at once, but you may wish to have them available.

Entering functions into Y= editor to be graphed

To enter functions to be graphed press the $\boxed{\text{Y=}}$ key.

Move the cursor to the particular Y= line you wish to use.

Note that the function *must* first be solved for y.

Enter two functions, say, $Y_1 = X + 2$ and $Y_2 = -X^2 - 2X + 1$.

Here are some tips for entering functions:
- Input the variable x with the $\boxed{\text{X,T,}\Theta,n}$ key
- Use the subtraction key $\boxed{-}$ between terms
- Use the negation key $\boxed{(-)}$ when the first term of a function is negative.

Press the $\boxed{\text{GRAPH}}$ key below the screen to view the graphs.

Turning functions on and off

The function or functions you wish to view must be turned on.

To do this, position the cursor over the = sign and
press [ENTER]. The equal sign will have a dark
rectangle over it.

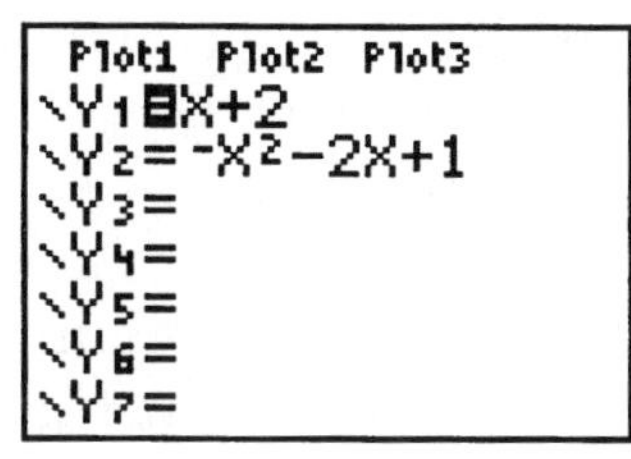

To turn a function off, again position the cursor over
the = sign and press [ENTER]. The equal sign will no
longer have the dark rectangle over it.

On the screen shown, $Y_1=X+2$ is on, and $Y_2=-X^2-2X+1$ is off.

Setting up a Window

The window setting defines the lowest and highest values of x and y that will
be displayed on the screen.

To set a window, press the [WINDOW] key below the screen.

    Xmin is the leftmost x-value shown on the screen.
    Xmax is the rightmost x-value shown on the screen.
    Xscl is the value of each tick-mark on the x-axis.

    Ymin is the lowest y-value shown on the screen.
    Ymax is the highest y-value shown on the screen.
    Yscl is the value of each tick-mark on the y-axis.

    Xres sets pixel resolution between 1 and 8.
    Use a higher Xres value when you want the graph to display more
    quickly. However, this will give rounded portions of a graph a more
    block-like appearance.

To change the window settings you may enter the values you wish, and then
press [GRAPH] or [TRACE] to see the graph in the new window.

You may also use the ZOOM key to select special zoom windows such as `Zoom Standard` and `Zoom Decimal`.

Move along a curve using TRACE

Press the TRACE key and use the left and right arrow keys to move along a curve. Use the up and down arrow keys to "jump" to another curve that is displayed.

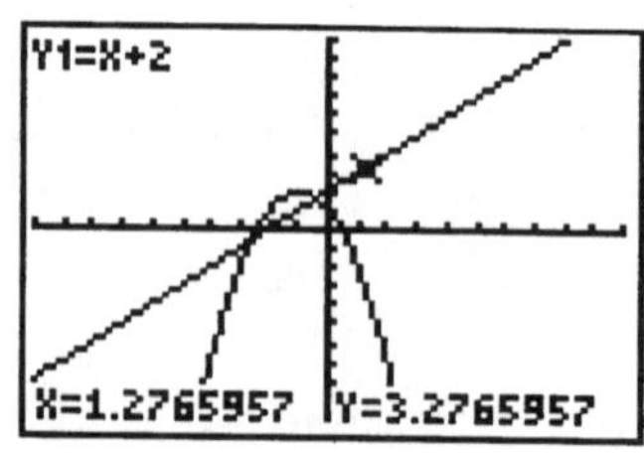

The x and y coordinates of the point the cursor is on are displayed at the bottom of the screen.

The equation of the curve you are on is displayed when the setting `ExprOn` is set. See directions for this setting below.

Expression On or Off

The TI-83 Plus has a graph format setting which lets you see the actual Y= equation on the screen as you trace along the curve.

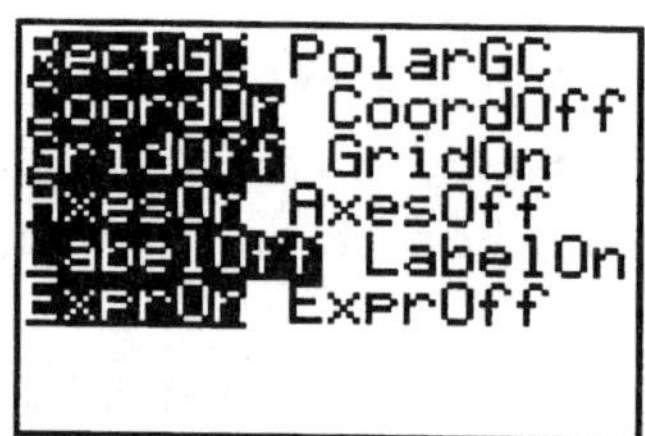

This setting is found by pressing 2nd ZOOM (`FORMAT`). Select `ExprOn` to display the Y= equation or `ExprOff` to turn this feature off.

The Standard Window

The standard window settings are
`Xmin=-10, Xmax=10, Xscl=1,`
`Ymin=-10, Ymax=10, Yscl=1, Xres=1.`

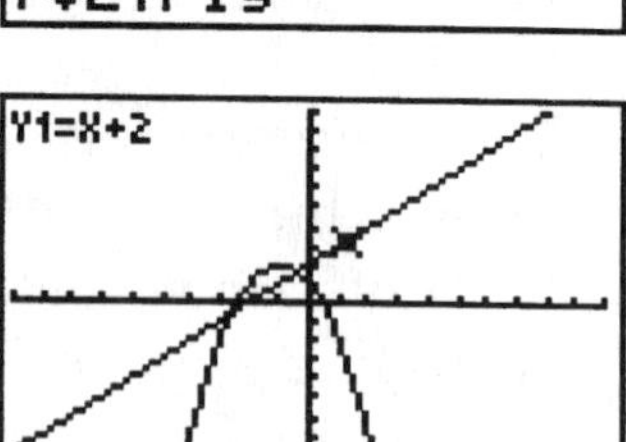

You can quickly set this window by pressing the ZOOM key and then the 6 key to select `ZOOM 6:ZStandard`.

Otherwise, press the ZOOM key, then move the cursor to `ZOOM 6:ZStandard`. Then press ENTER.

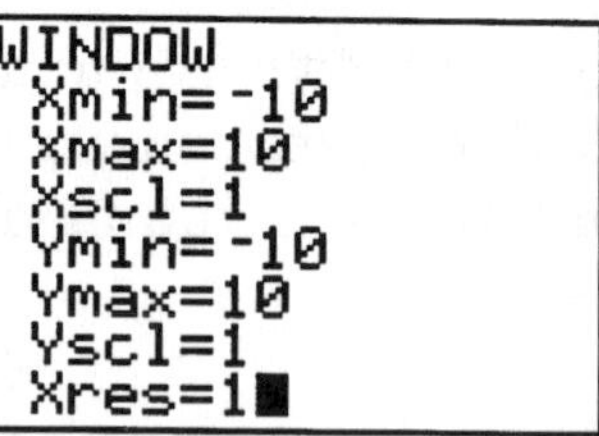

The graphs displayed are set in the standard window.

The Zoom Decimal (Friendly) Window

The Zoom Decimal window is sometimes called the friendly window.

The nice thing about this window is that when you trace a function, the coordinates of x and y shown on the graph screen have fewer decimal places.

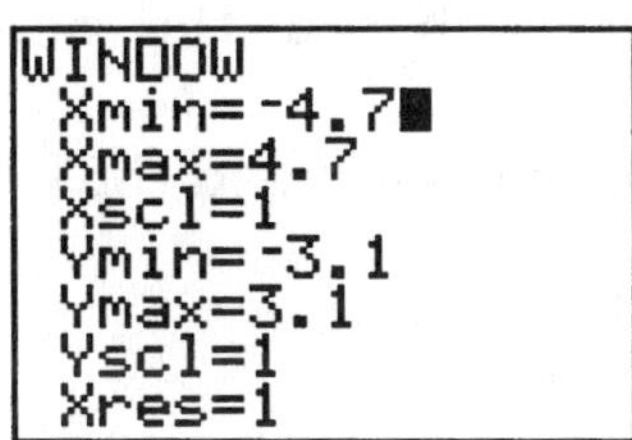

To set this window, press the [ZOOM] key and then the [4] key to select ZOOM4:Zdecimal.

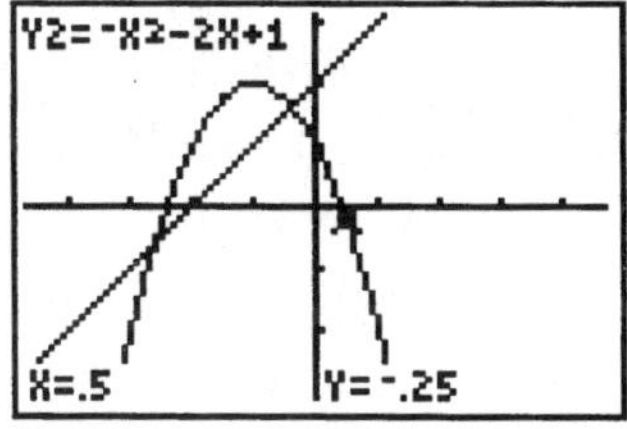

Otherwise, press the [ZOOM] key and then move the cursor to ZOOM4:ZDecimal. Then press [ENTER].

To maintain the features of Zoom Decimal reset Xmin and Xmax to multiples of −4.7 and 4.7. The values of Ymin and Ymax can be anything you want.

The graphs displayed are set in the Zoom Decimal window.

ZOOM IN and ZOOM OUT

Zoom In and Zoom Out allow you to zoom in or out around a point.

To zoom in, press the [ZOOM] key to put you on the ZOOM screen.

Move the cursor to 2:Zoom In and press [ENTER], or press the [2] key. This puts you on the graph screen and gives you a free-moving cursor.

Move the cursor to the point or area you wish to zoom in on with the arrow keys.

Press ENTER. The graph is redrawn with the point you chose as the center of the screen.

The first screen shows that we wish to zoom in at the + symbol with cursor coordinates (-1.1,1.3).

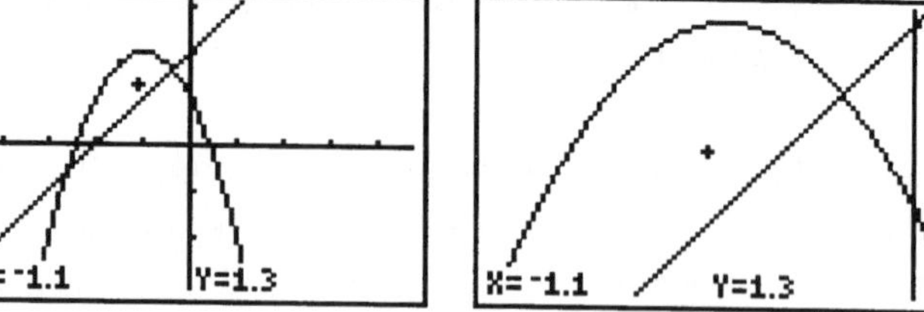

The second screen shows the graph after the zoom in.

Press WINDOW to see how the zoom changed the window of the graph.

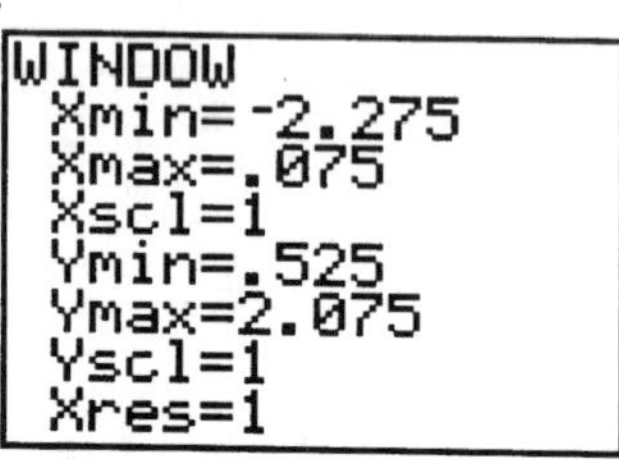

ZOOM OUT works the same way as ZOOM IN. To zoom out, press the ZOOM key to put you on the ZOOM screen.

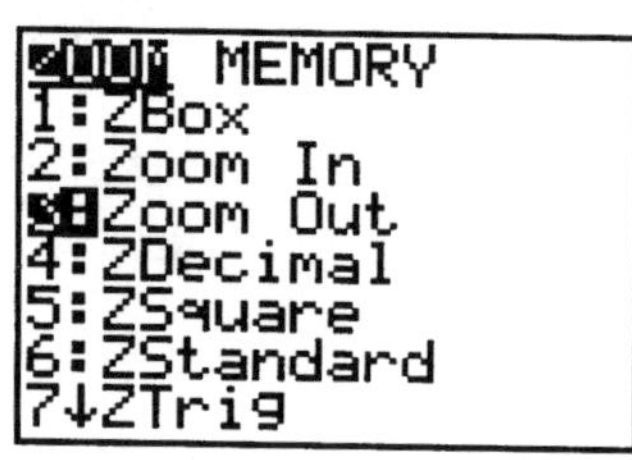

Move the cursor to 3:Zoom Out and press ENTER, or press the 3 key. This puts you on the graph screen and gives you a free-moving cursor.

Move the cursor to the point or area you wish to zoom out on with the arrow keys.

Press ENTER. The graph is redrawn with the point you chose as the center of the screen.

Press WINDOW to see how the zoom changed the window of the graph.

How to input a variety of functions in Y= :

Here are some sample functions and how to enter them to be graphed.

1. Enter a square root function such as $f(x) = \sqrt{x+2}$ as $Y_1=$ √(X+2). The √ symbol is found above the $\boxed{x^2}$ key. Note that all terms under the square root symbol must be enclosed in parentheses.

2. Enter a cube root function such as $f(x) = \sqrt[3]{x} - 1$ as $Y_1=$ ^{3}√(X)-1 using $\boxed{\text{MATH}}$ MATH 4: $\sqrt[3]{}$ $\boxed{\text{ENTER}}$.

3. Enter $f(x) = x^{2/5}$ as $Y_1=$ 5$\sqrt[x]{}$ X^2 using $\boxed{5}$ $\boxed{\text{MATH}}$ MATH 5: $\sqrt[x]{}$ $\boxed{\text{X,T,}\Theta\text{,}n}$ $\boxed{x^2}$ $\boxed{\text{ENTER}}$. First enter the root, 5, then the $\sqrt[x]{}$ symbol, then the radicand, X^2. Alternatively, enter $Y_1=$ X^(2/5). The $\boxed{\wedge}$ key is below the $\boxed{\text{CLEAR}}$ key.

4. Enter $f(x) = \dfrac{x+2}{x-1}$ as $Y_1=$ (X+2)/(X-1). Always enclose the numerator and denominator in parentheses when they have more than one term.

5. Enter $f(x) = 2 - |x-1|$ as $Y_1=$ 2 -abs(X-1) using $\boxed{\text{MATH}}$ NUM 1:abs(or using the catalog $\boxed{\text{2nd}}$ $\boxed{0}$ to locate abs(.

6. Enter $f(t) = 100\left(1 - \dfrac{12{,}196}{t^{4.478}}\right)$ as $Y_1=$ 100(1-(12196/X^4.478). Note that the t has been replaced with an X so that the calculator can understand the function.

BACK TO THE TEXT

Evaluate a function using 1.1 Example 3

Example 3 asks you to evaluate the function $f(x) = .03x^2 - .06x + 6$ for $x = 0, 1, \ldots, 10$.

Method 1 - Using Y_1:

Let's first see how to do this by using its function name, Y_1, on the Home Screen.

First enter the function $f(x) = .03x^2 - .06x + 6$ into the Y(X) menu as $Y_1 =$ or the first available Y= function.

Press [Y=], the first key below the screen, then enter $.03x^2 - .06x + 6$ as $Y_1 =$ or a different function in the Y= editor.

Use the [X,T,Θ,n] key to input X, and the [x²] key or [^] [2] to enter the exponent.

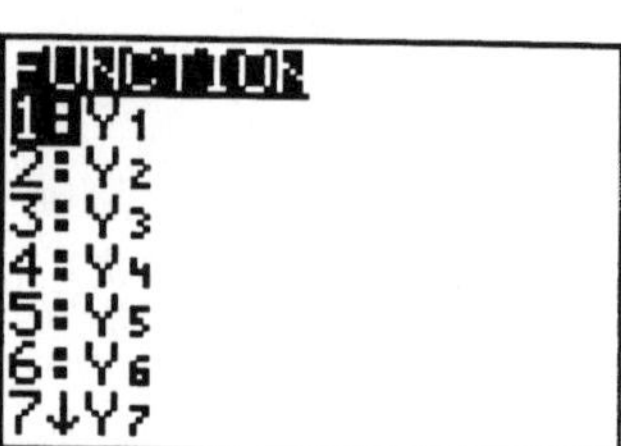

Press [2nd] [MODE] (QUIT) to go to the Home screen.

Next, press the [VARS] key.

Use the right arrow [▶] to move to Y-VARS, then press [1] to select 1: Function.

This brings you to another screen that has all the Y= names.

Move to 1: and press [ENTER], or press [1] to select 1:Y_1. This pastes Y_1 on the Home Screen.

Finally, type (0) next to the Y_1 to see $Y_1(0)$, and press $\boxed{\text{ENTER}}$ to see the result 6.

Repeat the line on the Home Screen by pressing $\boxed{\text{2nd}}$ $\boxed{\text{ENTER}}$ (ENTRY).

Use the left arrow to position the cursor over the 0. Press $\boxed{1}$ to overwrite it. Then press $\boxed{\text{ENTER}}$ to see the value 5.97.

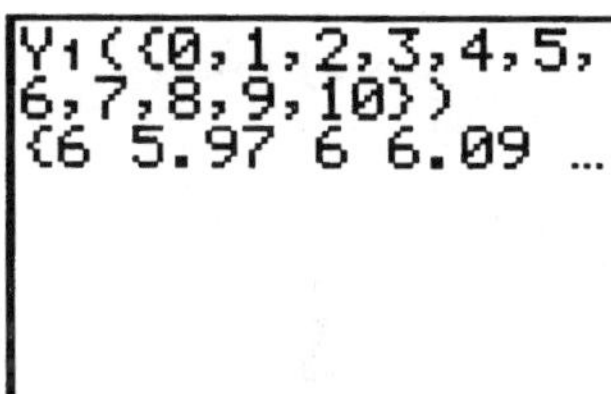

Repeat the line and change x again for $x=2,\ldots,10$. Press $\boxed{\text{ENTER}}$ to see the see each result. Record each result.

Method 2 - Using a List:

You can also use Y_1 on the Home Screen and a list of values to evaluate $f(x) = .03x^2 - .06x + 6$ for all the values of x from 1 through 10 on one line using braces $\{\,\}$.

Input Y_1 on the Home Screen as described above. Finish the command to read $Y_1(\{0,1,2,3,4,5,6,7,8,9,10\})$.

To input the left brace press $\boxed{\text{2nd}}$ $\boxed{(}$ ($\{$).
To input the right brace press $\boxed{\text{2nd}}$ $\boxed{)}$ ($\}$).
The comma $\boxed{,}$ is the key next to $\boxed{x^2}$.

Press $\boxed{\text{ENTER}}$ to see the values of $Y_1(0)$, $Y_1(1),\ldots,Y_1(10)$. The first number, 6, is $Y_1(0)$, the second number, 5.97, is $Y_1(1)$, and so on.

Use the right arrow to scroll through all the results.

Method 3 – Using a Table:

To evaluate $f(x) = .03x^2 - .06x + 6$ for $x=0,1,\ldots,10$ using the Table feature of the graph, be sure that f(x) is entered as Y_1.

Press $\boxed{\text{2nd}}$ $\boxed{\text{WINDOW}}$ to access TBLSET.

Set Tbl Start to 0; ∆ Tbl to 1; Indpnt: to Auto, Depend: Auto.

Press [ENTER] after each selection.

Next, press [2nd] [GRAPH] (TABLE) to see the table.

Use the down arrow to scroll through the table.

Note that, even though you tell the calculator to start at x=0, you can use the up and down arrows to scroll to any value of x you wish.

Use the right and left arrows to highlight the value you wish to focus on. The highlighted value is shown at the bottom of the screen.

You can also use the TBLSET and select Indpnt: Ask. This setting requires you to input the values you want to see.

Input the values x=0, 1, . . . , 10 one at a time.

Press [ENTER] each time to see the corresponding value of f(x).

Note that after x=6, the calculator does not scroll to a new line. The calculator overwrites the last line.

Press [2nd] [MODE] (QUIT) to return to the Home screen.

Method 4 – Evaluate a Function Using Its Graph and CALC 1:value:
Another way to evaluate a function on the graph screen uses CALC, a feature
of the graph menu.

Enter $Y_1 = .03x^2 - .06x + 6$ to be graphed. Press [2nd] [TRACE] (CALC) to access
CALC.

Press [1] to select CALCULATE 1:value.

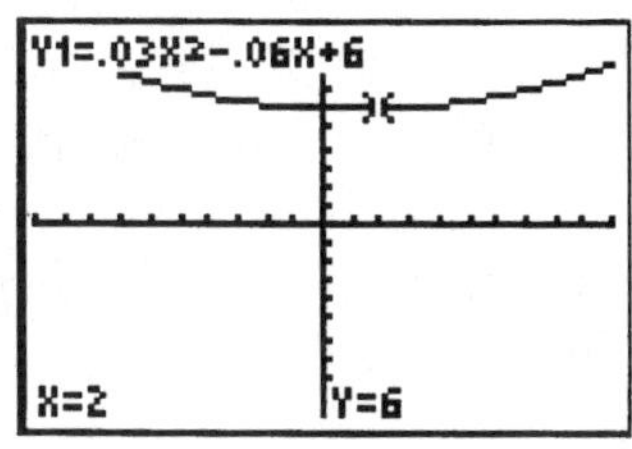

The graph of Y_1 is shown on the screen, and X= is
flashing at the bottom of the screen.

Enter the value of x you wish to evaluate. The values
of x and y appear at the bottom of the screen, and the cursor jumps to the
corresponding point. Record the result.

You may immediately enter the next value of x you wish to evaluate. Record
the result.

When more than one graph is displayed, you can see the value of y for a given
x on another graph by using the up or down arrow.

Note that CALCULATE 1:value works only in the Xmin to Xmax interval for
the current window displayed. If you wish to evaluate the function outside
the window, change the value of Xmin and/or Xmax to include the value.

Test a value in an inequality
In the Graphing Calculator presentation of Example 4 you are given the
function
U =(X≤10)(76-3.3X)+(x>10)(18+2.5x).

When a value of X, say 2, is less than or equal to 10, the expression (X≤10),
will evaluate to 1 while the expression (X>10) will evaluate to 0. The TI-83
Plus has been programmed to assign a true statement is assigned a 1 and a
false statement to 0.

Similarly, when X is greater than 10, say 11, (X≤10) will evaluate to 0, and
(X>10) will evaluate to 1.

To see this on the calculator, input 2≤10 and press ENTER to see the result 1.
Remember that 1 means true.

To input ≤, press 2nd MATH (TEST). Move the
cursor to 6:≤ and press ENTER , or press 6 to
paste ≤ on the Home screen.

Then input 2>10, and press ENTER to see the
result 0. The result 0 means false.

```
2≤10
              1
2>10
              0
11≤10
              0
11>10
```

To input >, press 2nd MATH (TEST). Move the cursor to 3:> and press
ENTER , or press 3 to put > on the Home screen.

Repeat for 11≤10 and 11>10 to see the results 0 and 1 respectively.
Note that the last result of 1 for 11>10 scrolls to the next screen, and is not
shown here.

To understand how each
term evaluates, let's input
each term of U separately
when x=2 and x=11 as
shown.

```
(2≤10)(76−3.3*2)     (11≤10)(76−3.3*1
              69.4   1)
(2>10)(18+2.5*2)                     0
                     (11>10)(18+2.5*1
              0      1)
                                  45.5
```

Notice that when x=2, the first term of U gives 69.4 while the second term of
U is 0.

Also, note that when x=11, the first term of U is 0 and the second term of U is
45.5.

Finally, input U entirely when x=2 and x=11 as shown.

This agrees with the above
results that
when X=2, U=69.4, and
that when X=11, U=45.5.

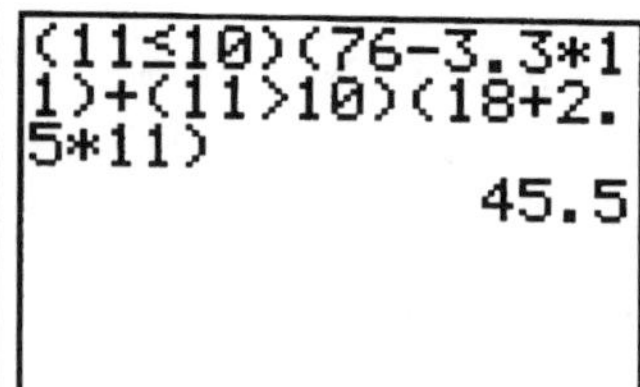

Input piece-wise functions to be graphed using 1.1 Example 4

In Example 4 of section 1.1 you may wish to graph

$$U(t) = \begin{cases} 76 - 3.3t & \text{if } 1 \le t \le 10 \\ 18 + 2.5t & \text{if } 10 < t \le 14 \end{cases}.$$

This function is defined differently over two intervals and is a little tricky to input into your calculator.

First, change the way the graph screen displays a graph so that it makes a separate dot for each point it graphs, rather than joining points with a line segment.

To do this, press the MODE key. Move the cursor to Dot on the 5th line and press ENTER.

Press CLEAR or 2nd MODE (QUIT) to leave the MODE screen.

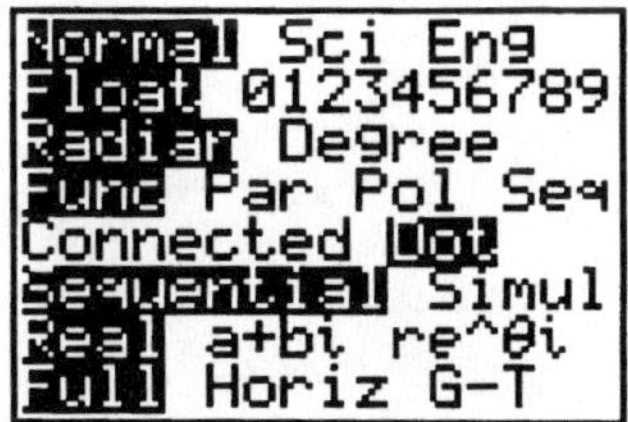

Next, press the Y= key, and either clear Y_1 or turn it off. Then input the new function as Y_2 or some other function.

To turn off a function so that it does not graph, position the cursor over the equal sign, and press ENTER.

Next, input the function $U(t) = \begin{cases} 76 - 3.3t & \text{if } 1 \le t \le 10 \\ 18 + 2.5t & \text{if } 10 < t \le 14 \end{cases}$

into the Y= menu as

Y_1 =(1≤x and x≤10)(76-3.3x)+(10<x and x≤14)(18+2.5x).

Use the Test menu 2nd MATH (TEST) to find the inequality symbols.

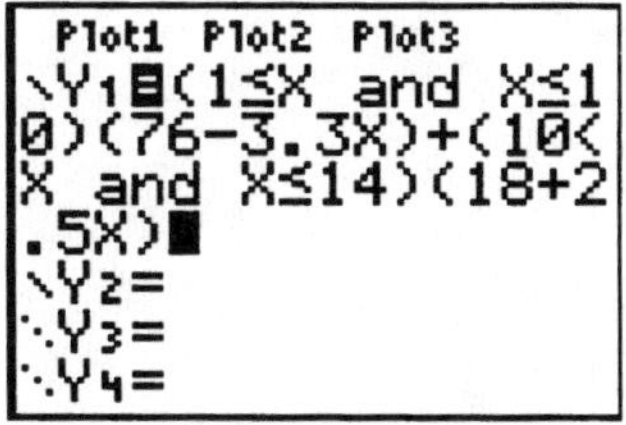

Use the word a n d to make two inequalities for
$1 \leq x \leq 10$ and for $10 < x \leq 14$. It is necessary to make two inequalities for each
because the TI-83 Plus does not understand a compound inequality.

Find a n d in the Test menu 2nd MATH LOGIC 1:and
or in the catalog 2nd 0. Move the cursor to the
word a n d. Press ENTER .

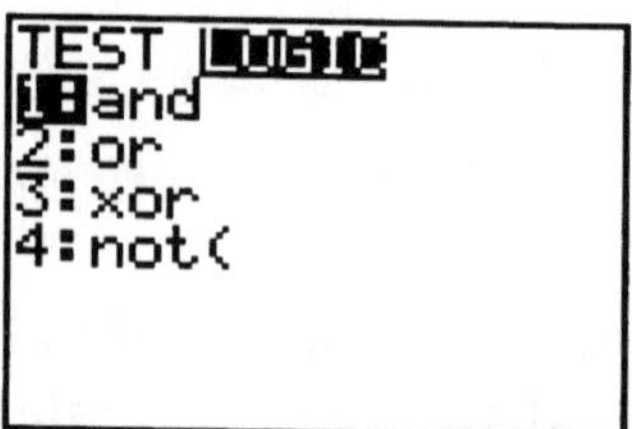

Next, adjust the window to
Xmin= -1, Xmax=15, Xscl=1,
Ymin=30,   Ymax=80, Yscl=5.

Press GRAPH to see the graph of this piece-wise
function.

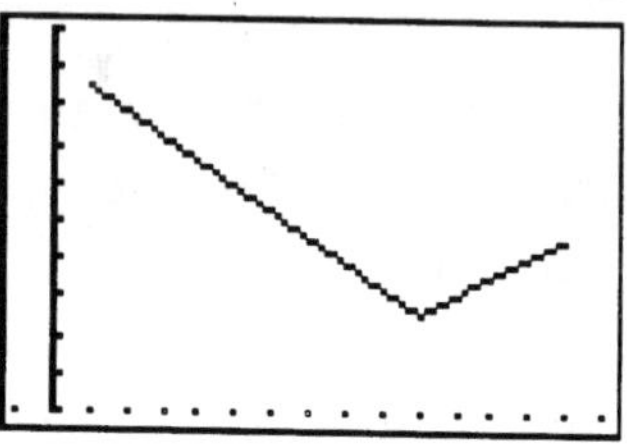

1.2 Functions from the Graphical Viewpoint - TI-83 Plus

In this section you will learn how to plot individual points on the graph
screen, use Stat Plot to format the graph of data points, enter functions to be
graphed, turn them on and off, set windows, and graph a variety of functions.
You will also see how to draw a vertical line to demonstrate the Vertical Line
Test.

Plot individual points on the graph screen

Plotting individual points on the graph screen is possible using the STAT
menu of the calculator and defining and selecting Plot 1, Plot 2, or
Plot 3.

Let's use 1.2 Example 2 *Graph of a Function: Web Site Revenue* to
demonstrate how this is done.

Example 2 states: *The monthly revenue R from users logging onto your
computer game site depends on the monthly access fee p you charge according
to the formula* $R(p) = -5{,}600p^2 + 14{,}000p$, *where* $(0 \le p \le 2.5)$ *and* (*R and p are
in dollars.*)

*Sketch the graph of R: Find the access fee that will result in the largest
monthly revenue.*

The solution requires that you plot several points. Here is how to do this on
the graphing calculator.

Plot the following points for $R(p) = -5{,}600p^2 + 14{,}000p$ $(0 \le p \le 2.5)$

p	0	0.5	1	1.5	2	2.5
$R(p) = -5{,}600p^2 + 14{,}000p$	0	5,600	8,400	8,400	5,600	0

Set the Window

First set the window.
Press the [WINDOW] key.

Set the window elements to
$Xmin=0, Xmax=2.5, Xscl=.5,$
$Ymin=0, Ymax=10000, Yscl=1000.$

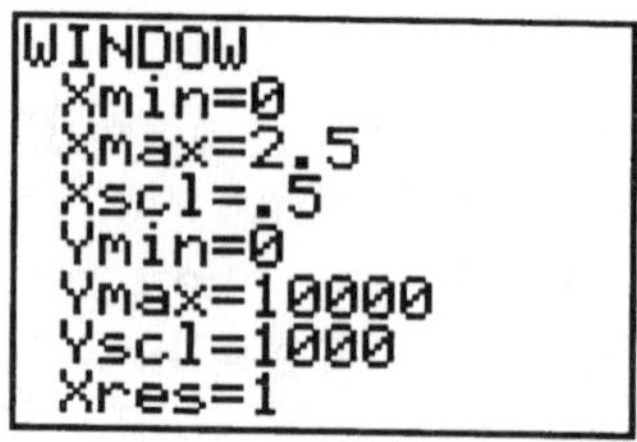

Press [ENTER] after each entry.

Enter the Data Points

Next, press the [STAT] key. The word EDIT at the
top of the screen should be highlighted (dark).

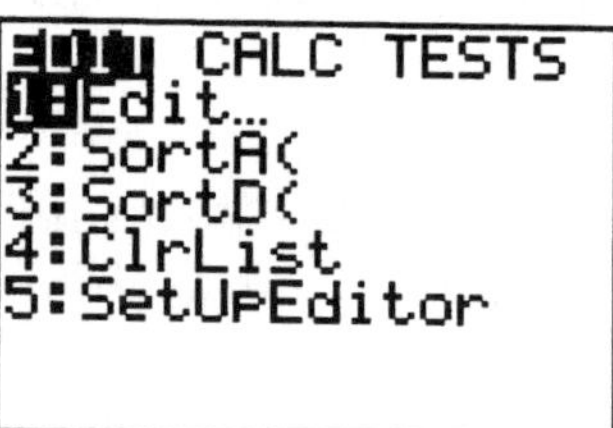

Press [1] to select 1:Edit or move the cursor to
1:Edit and press [ENTER]. This brings you to the
screen where you can enter the data.

If you wish to clear the lists, go back to STAT EDIT, and select 4:
ClrList. This pastes ClrList on the Home Screen.

Press [2nd] [1] (L_1) [ENTER] to clear L_1. You should see
the word Done.

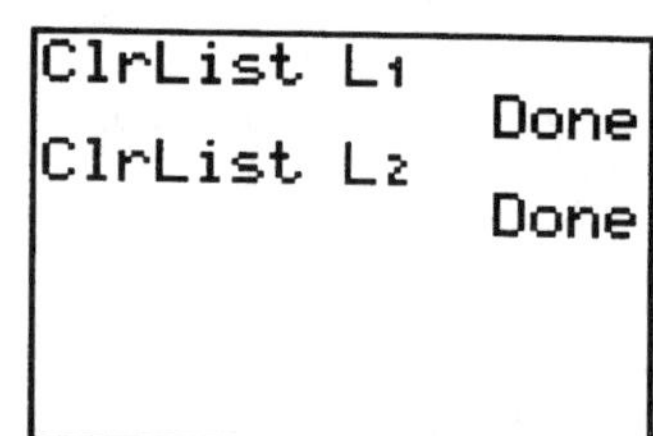

Repeat the line by pressing [2nd] [ENTER] (ENTRY),
and move the cursor over L_1. Press [2nd] [2] (L_2) to
clear L_2.

Another way to clear entries from a list is to overwrite the elements of the list
or delete any elements you want using the [DEL] key.

Note that the [CLEAR] key will not erase a highlighted value.

However, a list can be cleared in the STAT menu by pressing the up arrow
until the top of the list is highlighted. Pressing [CLEAR] [ENTER] in this position
clears the contents of the entire list.

If you press the DEL key when the column heading is highlighted, it will remove the list from the STAT editor, but the list will still possess its contents. The list name can be returned to the editor with 2nd DEL (INS) command and typing in the list name.

Enter the p-values in list L_1.

Enter the R(p)-values in L_2.

Press ENTER after each entry.

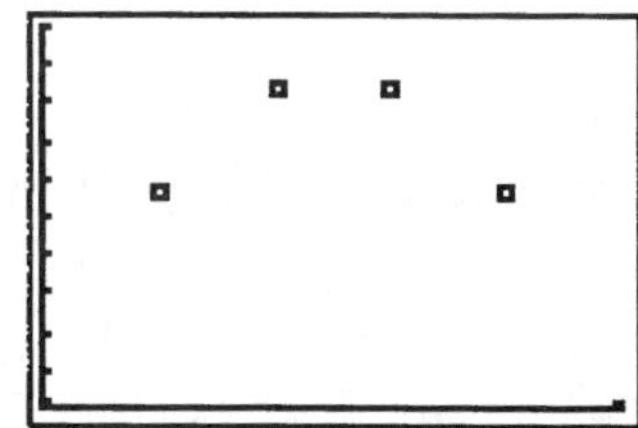

When you are finished, go back to the Y= menu and turn off all functions, namely the ones that have a dark rectangle over the equal sign.

To do this, move the cursor over the equal sign of the function you wish to turn off.

Press ENTER to remove the highlighted (dark) rectangle over the equal sign.

Turn on a Plot

You must now turn on one of the Plots. Let's turn on Plot 1.

In the Y= menu use the up arrow to go to Plot 1. Press ENTER to highlight it.

Press GRAPH to see the points.

Adjust the window if you wish to see the points (0,0) and (2.5,0) more clearly.

If you see something else, like a histogram, you must format the Plot using STAT PLOT as described below.

Use STAT PLOT to format the graph of data points.

You may change the type of plot, redefine the lists assigned to the x and y axes, and select a Mark for each plot you wish to graph.

To do this, press [2nd] [Y=] to access STAT PLOT. To select Plot 1, Plot 2, or Plot 3, move the cursor to the plot you want. Press [ENTER].
Next, make your selections.

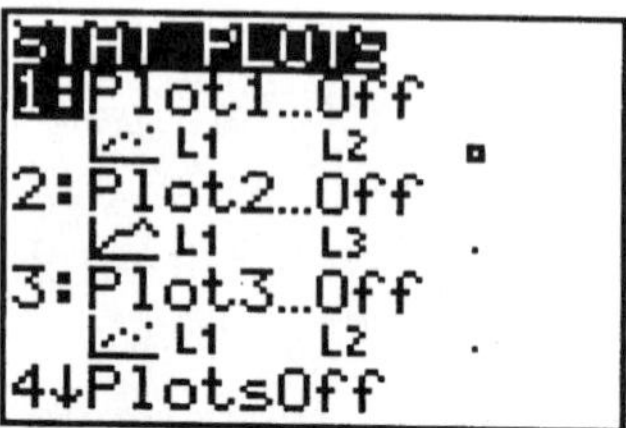

Press [ENTER] after each selection.

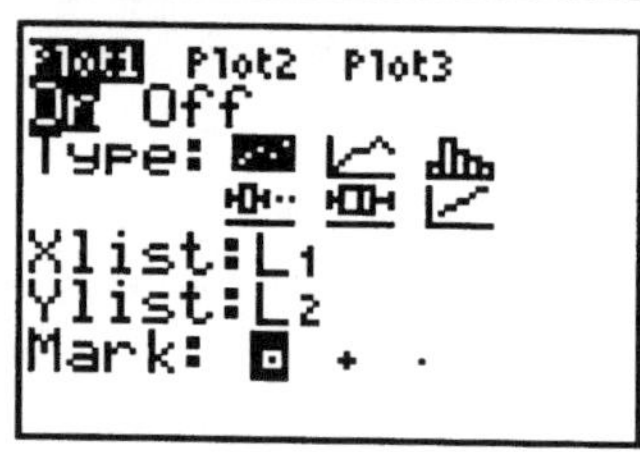

  Select ON
  Type: first one for a scatter plot – individual points
  XList: L_1 or whichever list has the x-data
  YList: L_2 or whichever list has the y-data
  Mark: any one you wish

Press [GRAPH] to see the scatter plot.

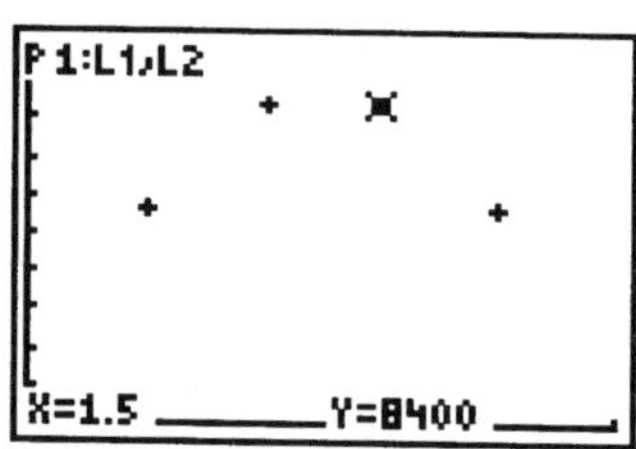

Note: You may also see other Y= functions on the same screen. If you do not wish to see these, turn them off.

Use [TRACE] to move from point to point. The x and y coordinates of each point are displayed at the bottom of the screen.

Note that P1:L1, L2 is displayed in the upper left hand corner of the screen. This tells you that you are on Plot 1, the x's are from list L_1, and the y's are from list L_2.

Graph the Function Through its Data Points

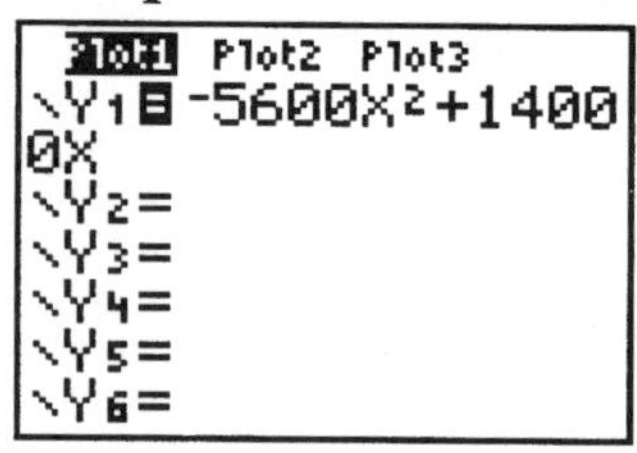

Enter
$R(p) = -5{,}600p^2 + 14{,}000p$ as Y_1
to show the equation going through the points.

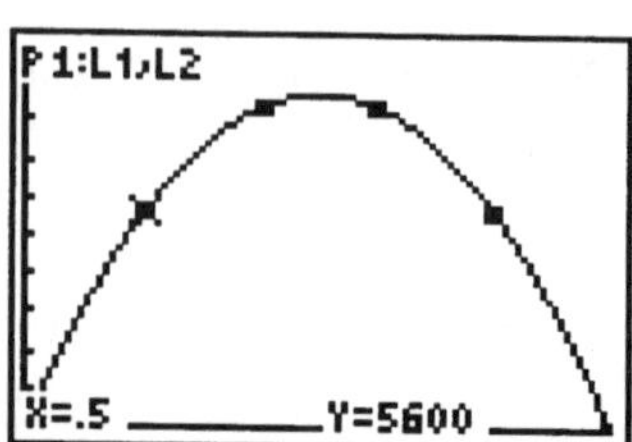

Notice that you enter the variable p as X using the $\boxed{\text{X,T,}\Theta\text{,}n}$ key, and enter the exponent with the $\boxed{x^2}$ key.

Press $\boxed{\text{GRAPH}}$ to see the graph of the equation through the data points.

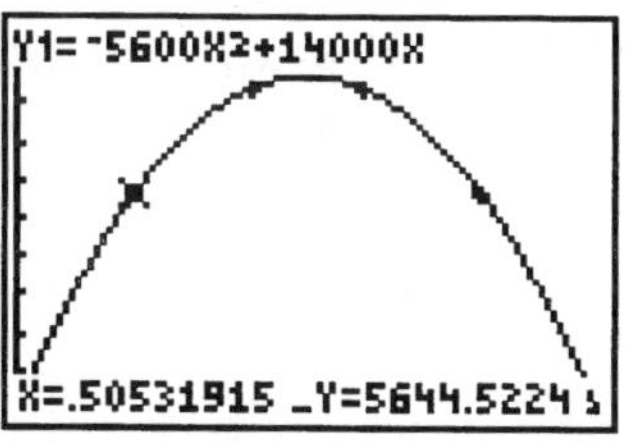

Use the up or down arrow to toggle between the function $Y_1=-5600x^2+14000x$ and the data points.

Use a vertical line to demonstrate the Vertical Line Test

The Vertical Line Test helps you to decide whether or not a graph is a function. On the calculator, it is possible to draw a vertical line using the DRAW menu and then move it through the graph using the right and left arrow keys.

First, enter or turn on a function in the Y= menu, say $R(p) = -5{,}600p^2 + 14{,}000p$ of Example 2 that was done earlier in this section.

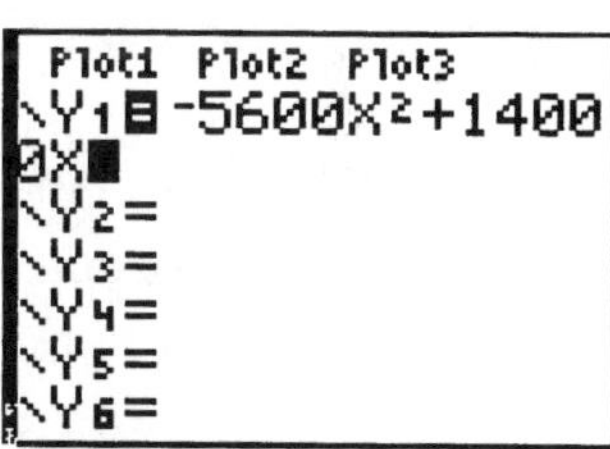

Turn off the Plot.

Use the same window as shown.

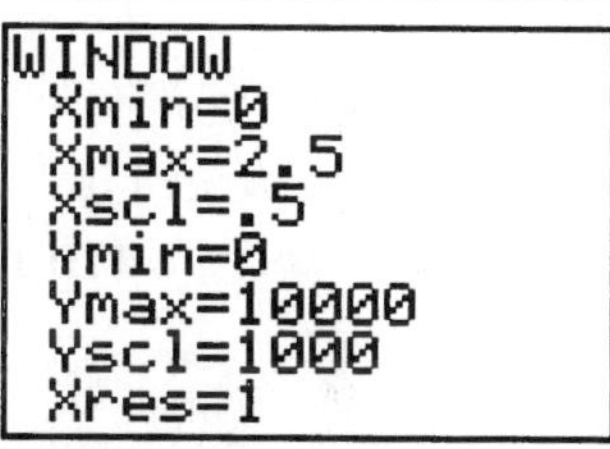

Now press $\boxed{\text{2nd}}$ $\boxed{\text{PRGM}}$ (DRAW) to access the DRAW menu and select DRAW 4:Vertical.

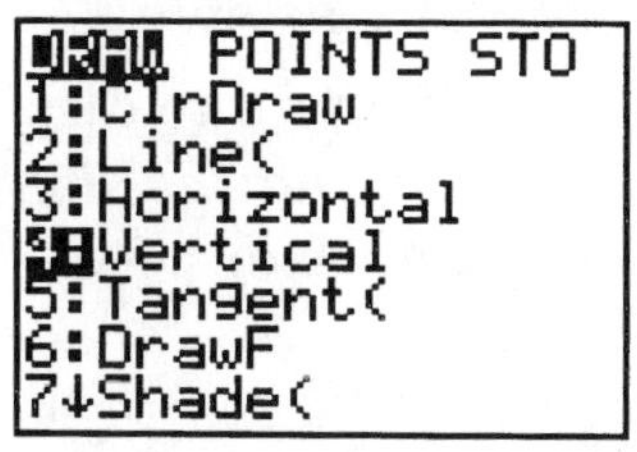

This superimposes a vertical line on the graph of $R(p) = -5{,}600p^2 + 14{,}000p$.

Use the right and left arrows to move the vertical line across the graph.

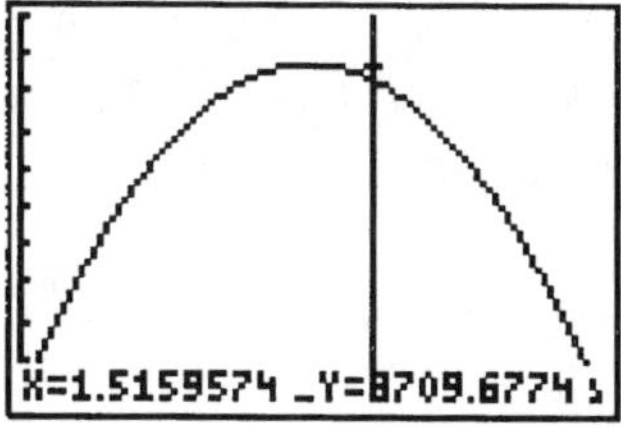

1.3 Linear Functions - TI-83 Plus

In this section, you will see how to square up the graphs using ZOOM
5:ZSquare, evaluate the slope of a linear function on the Home Screen, and
express ANS as a fraction. Lastly, you will see how to fit a linear equation to
data.

Square up the graphs using ZOOM 5:ZSquare

Sometimes it is helpful to see a graph as if it
were drawn on graph paper. This is called
squaring up the graph.

To do this for a selected graph, press the ZOOM
key below the screen. Select 5:ZSquare and
press ENTER . The graph is redrawn and squared up.

Evaluate slope on the Home screen:

Example 3b of this section asks you to *find the slope of the straight line
through the points (1,2) and (3, -1).*

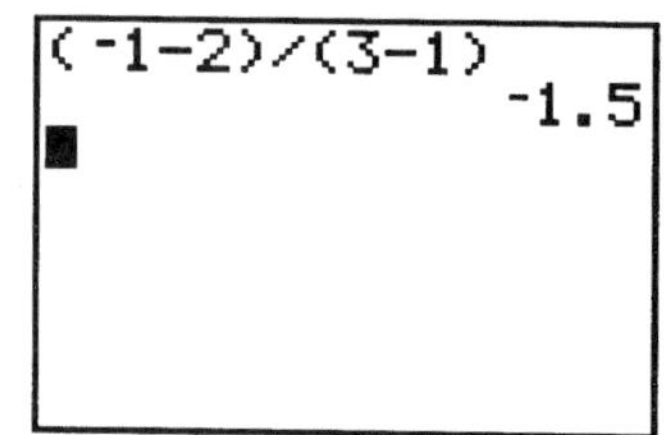

To do this, enter (-1-2) / (3-1) on the Home
Screen. Be sure to enter the subtraction ⊟ sign
between terms and the negation ⊡ sign for
negative numbers such as -1. Press ENTER to
see -1.5.

To see your answer ANS as a fraction:

It is often helpful to display your answer as a fraction. Let's do this for
Example 3b *find the slope of the straight line through the points (1,2) and (3, -1),* whose answer is −1.5.

Press the MATH key to see the MATH menu.

Be sure that MATH at the top of the screen is
highlighted (dark).

Press 1 to select 1:→Frac.   This pastes →Frac on the Home screen.

Press $\boxed{\text{ENTER}}$ to see $-3/2$.

Or, from the MATH menu, move the cursor to
1:$\rightarrow$Frac and press $\boxed{\text{ENTER}}$ to pastes $\rightarrow$Frac on
the Home screen. Press $\boxed{\text{ENTER}}$ again to see $-3/2$.

Fit a linear equation to data

In Example 3b you are asked to *find the equation of the straight line through
the points (1,2) and (3, -1).*

To find the equation, enter the two points into
$\boxed{\text{STAT}}$ Edit. (See 1.2 Plot Individual Points.)

Put the x coordinates into L_1 and the y coordinates
into L_2.

You may need to clear the lists using $\boxed{\text{STAT}}$ Edit 4:ClrList or $\boxed{\text{DEL}}$ to
delete the values in the lists.   You can also clear lists that are in the STAT
editor by moving to the top of the desired list to highlight the list name.
Press $\boxed{\text{CLEAR}}$ $\boxed{\text{ENTER}}$.

Next, press the $\boxed{\text{STAT}}$ key and use the right arrow
to go to CALC.

Press $\boxed{4}$ to select 4:LinReg(ax+b).   This puts
LinReg(ax+b) on the Home screen.

Press $\boxed{\text{ENTER}}$ to see a=-1.5, b=3.5.

Or, you can go to $\boxed{\text{STAT}}$ CALC, then move the
cursor down to 4:LinReg(ax+b) and press $\boxed{\text{ENTER}}$
to paste LinReg(ax+b) on the Home screen.
Press $\boxed{\text{ENTER}}$ to see a=-1.5, b=3.5.

Hence, the equation of the line through the points (1,2) and (3, -1) using a
and b is

$$y = -1.5x + 3.5 \quad \text{or} \quad y = -\frac{3}{2}x + \frac{7}{2}.$$

1.4 Linear Models - TI-83 Plus

There are no new calculator skills in this section.

1.5 Linear Regression - TI-83 Plus

This section asks you to find the correlation coefficient of a regression equation. To do this you must turn the diagnostics on using `DiagnosticOn` as described below.

Set DiagnosticOn to find the correlation coefficient.

In the previous section you found the linear regression equation for two points. Recall that the LinReg screen only showed you the values of a and b.

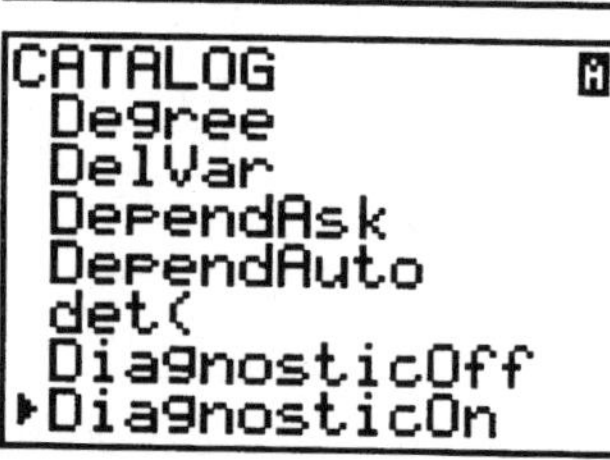

In this section you are asked to also find the correlation coefficient, `r`.   The TI-83 Plus will give you this information, but you must turn on this capability by using the setting, `DiagnosticOn`.

To turn on the diagnostics so that the correlation coefficient is displayed with the linear regression equation, press [2nd] [0] to access `CATALOG`.

Press [x⁻¹] to jump to the D's and use the down arrow to find `DiagnosticOn`.

Press [ENTER] twice until you see the word `Done` on

the Home Screen.

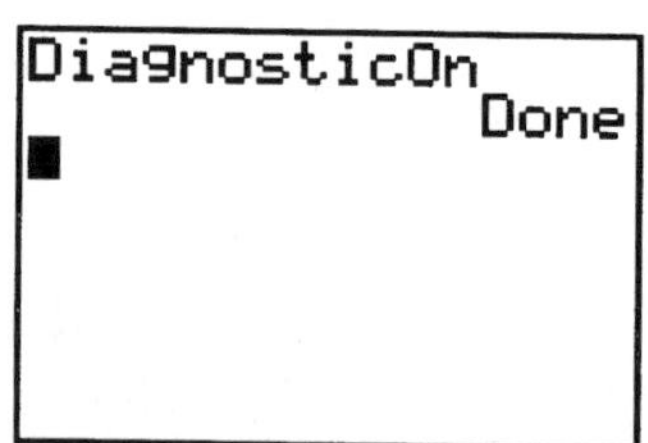

When you do this for the linear regression equation for two points of the previous section, you get the screen with the additional values of r and r^2 as shown.

Let's first use 1.5 Example 2 - *New Home Sales to find the linear model using the data on new home sales* to input the data. Next, with `DiagnosticOn`, use Example 3 *Correlation Coefficient to find the correlation coefficient for the linear model using the data on new home sales.*

Selling Price (Thousands of $)	160	180	200	220	240	260	280
Number sold	126	103	82	75	82	40	20

First you must input the selling prices into L_1 and the number of homes sold into L_2.

Press `STAT`. Select `EDIT 1:Edit` to enter the values. This screen is called the `STAT` editor.

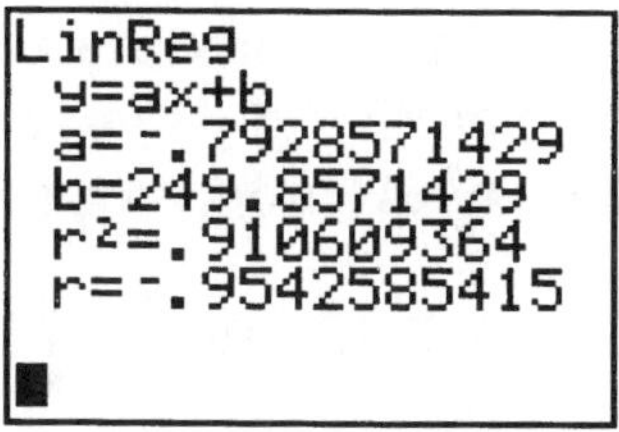

Next, find the Linear Regression equation. Go back to `STAT` `CALC 4:LinReg(ax+b)`.

Press `ENTER` to see `a`, `b`, r^2, and `r`.

Rounding `a` to three decimal places, we get `a ≈ -0.793`. Rounding `b` to the nearest integer, we get `b ≈ 250`.

Hence, the least squares line is `y=-0.793x + 250`.

Finally, rounding `r` to four decimal places we get `r ≈ 0.9543`.

You're the Expert – Modeling Spending on Internet Advertising - TI-83 Plus

This section asks you to find a quadratic model for a set of data points and to find the correlation coefficient r for the quadratic model. Lastly, you are asked to input the regression equation in the Y= menu to see its graph.

First, plot the data points on the graph screen.
Use the data points given in this section to find the linear and quadratic models as well as the correlation coefficient for each model.

Year	1995	1996	1997	1998	1999	2000	2001
Spending on Advertising ($ Billion)	0	0.3	0.8	1.9	3	4.3	5.8

Plot the data points on the graph screen

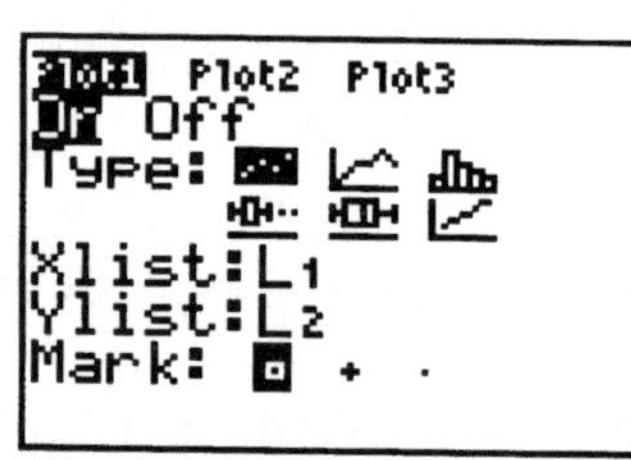

First input the data points into L_1 and L_2. Let's use the year 1995 as year 0, 1996 as year 1, etc. Input the years into L_1 and the spending into L_2.

Press STAT. Select EDIT 1:Edit to enter the values.

Press 2nd Y= (STAT PLOT) to set up Plot 1 as shown.

Set the window to Xmin=-1, Xmax=8, Xscl=1, Ymin=-1, Ymax=7, Yscl=1.

Press GRAPH to see the data points. Press TRACE to see P1 : L_1, L_2 as shown.

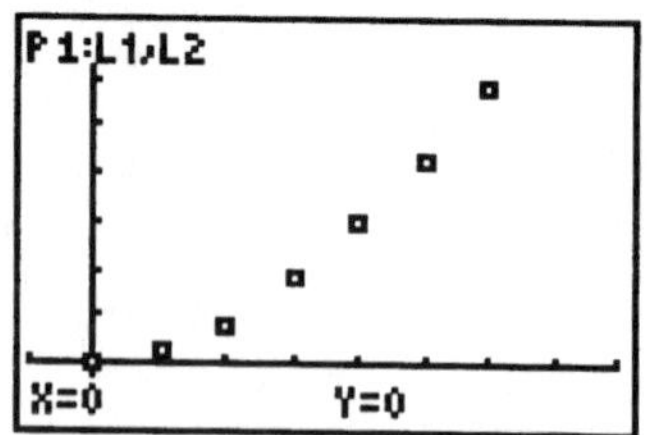

Find the linear model

Next, find the Linear Regression equation.

Go back to $\boxed{\text{STAT}}$ CALC 4:LinReg(ax+b). Press $\boxed{\text{ENTER}}$ to see a, b, r^2 and r. Rounding to four decimal places we find that a=0.9857, b=-0.6571, and r=0.9781.

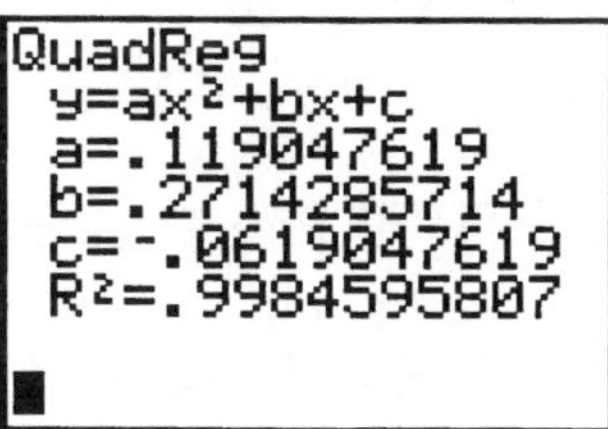

Hence, the linear regression model is y=0.9857x - 0.6581, and the correlation coefficient r is 0.9781.

Find a quadratic model

To find the quadratic model for the "You're the Expert" example, note that we already have the data points in L_1 and L_2.

Press $\boxed{\text{STAT}}$. Move to CALC and select 5:QuadReg. This puts QuadReg on the Home screen.

Press $\boxed{\text{ENTER}}$ to obtain the coefficients a, b, and c of the quadratic model and R^2.

Find the correlation coefficient r from R^2

We are interested in the correlation coefficient r, not R^2. We can calculate r on the Home Screen using $\boxed{\text{VARS}}$.

On the Home Screen press $\boxed{\text{2nd}}$ $\boxed{x^{-1}}$ to access $\sqrt{\ }$.

Press $\boxed{\text{VARS}}$. Select VARS 5:Statistics to access the statistics variables.

Move the right arrow twice to EQ to see the equation variables. Scroll down to the bottom of the list. Select 9: R^2.

Press ENTER to put R^2 on the Home screen. Input the right parentheses.

Press ENTER again to see $r = 0.9992294935$ on the Home screen. This number rounded to four decimal places is $r = .9992$.

Input the Regression Equation in the Y= menu

We would now like to see the graph of the quadratic regression equation for the "You're the Expert" example.

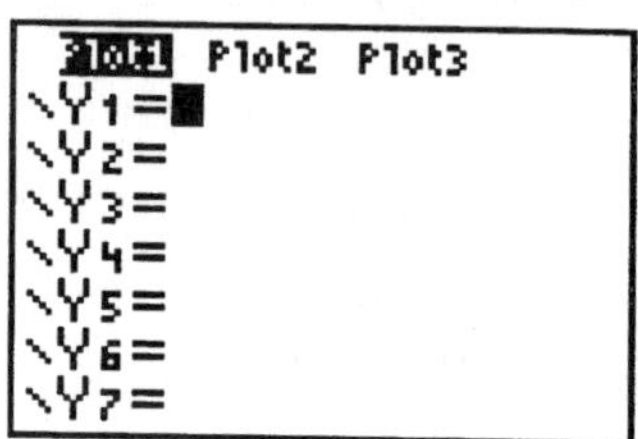

First, be sure that Plot 1 is turned on and formatted as described earlier.

In the Y= menu put the cursor next to the Y= equation where you wish to put the regression equation, say into Y_1.

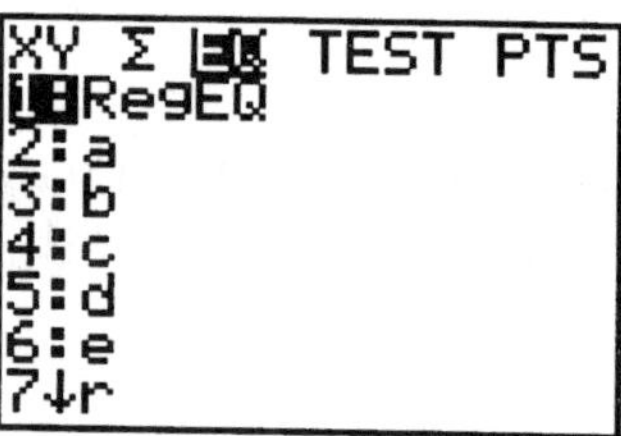

Press VARS 5:Statistics to access the statistics variables. Move the right arrow twice to EQ to see the equation variables. Select 1: RegEQ.

Press ENTER to see the equation
$$Y_1 = .11904761904762x^2 + .27142857142857x + -.0619047619048 .$$

Move the up arrow to Plot 1 and press ENTER to turn it on. Set the window to
Xmin=-1, Xmax=8, Xscl=1,
Ymin=-1, Ymax=7, Yscl=1.

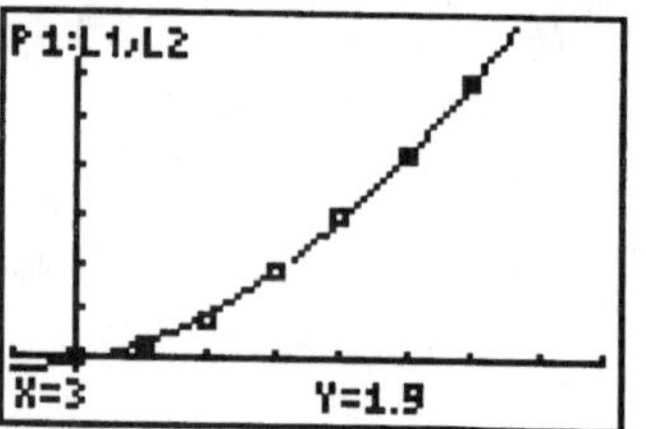

Press GRAPH to see the graph of the points and the quadratic model. Use TRACE to move through the points. Use the up arrow to move to TRACE along the quadratic model.

CHAPTER 2

SYSTEMS OF LINEAR EQUATIONS AND MATRICES

TI-83 PLUS

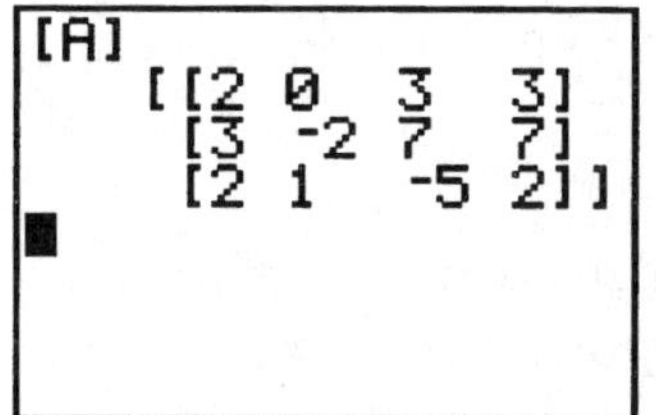

2.1 Systems of Two Linear Equations in Two Unknowns TI-83 Plus

In this section you are asked to find the point of intersection of two graphs. Let's examine three ways to find the point of intersection: TRACE and ZOOM, ZBox, CALC. We will use 2.1 Example 1 – *Two Ways of Solving a System.* The two equations of the system are

$$x + y = 3$$
$$x - y = 1$$

First we must solve each equation for y to get the new system that we input as

$$Y_1 = -X+3$$
$$Y_2 = X-1$$

Find the point of intersection Using TRACE and ZOOM

First enter the two equations $Y_1 = -X+3$ and $Y_2 = X-1$ into the Y= menu.

Next, set the window by pressing the WINDOW key. Set Xmin=-4, Xmax=4, Xscl=1. Press the GRAPH key to see the graph of the two functions.

Press the TRACE key and use the arrow keys to move the cursor as close as possible to the intersection. Use Zoom In, Zoom Out, or ZBox and TRACE to get closer to the intersection.

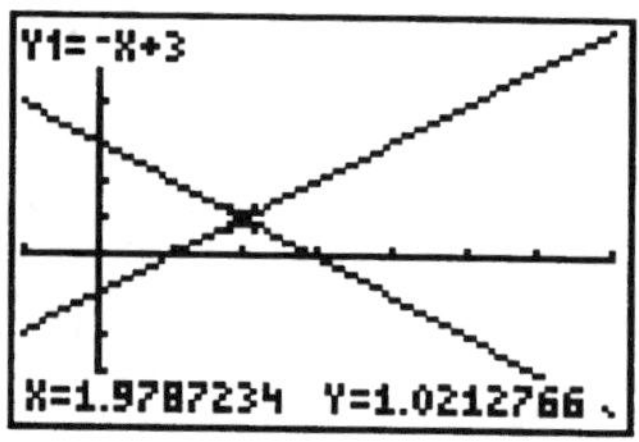

The TI-83 Plus has a graph format setting which lets you see the actual Y= equation on the screen as you trace along the curve. This setting is found by pressing 2nd ZOOM (FORMAT). Select ExprOn to display the Y= equation or ExprOff to turn this feature off.

Using ZBox and Trace

ZBox lets you create a box around interesting parts of a graph such as the points of intersection.

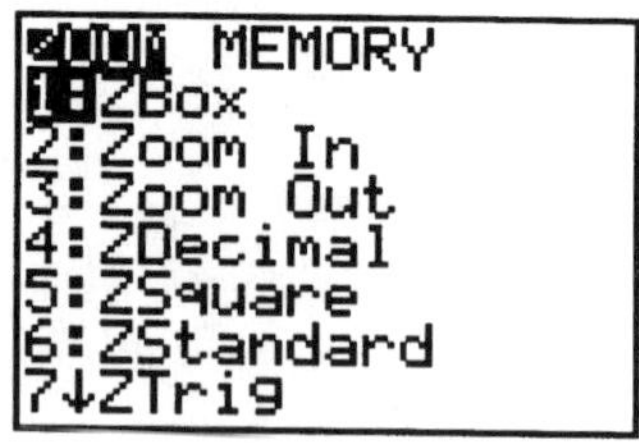

To use ZBox press ZOOM. Press 1 to select 1:ZBox, or move the cursor to 1:ZBox and press ENTER. The graph is displayed and there is a free moving cursor.

Move the cursor using the arrow keys close to the upper left of the point of intersection and press ENTER .

Move the cursor to the right and down to see a box forming around the point of intersection.

Press ENTER somewhere to the lower right of the point of intersection. The graph will be redrawn with a new window defined by the box you created.

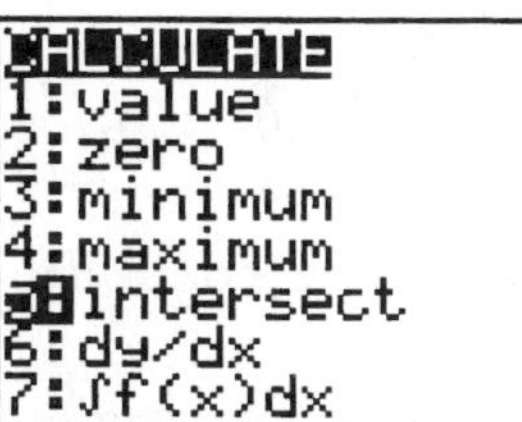

Then use TRACE again to get closer to the point of intersection. Repeat the process if needed.

Find the intersection using CALC

Yet another way to find the intersection is to use a feature of the graph menu called CALC.

Enter the two equations $Y_1 = -X+3$ and $Y_2 = X-1$ into the Y= menu.

Press 2nd TRACE (CALC) below the graph screen. Select 5:intersect.

You are asked which for the first curve. Notice the number of the equation or the actual equation on the TI-83 Plus.

Press ENTER to select the first curve. Press ENTER again to select the second curve. When asked for a GUESS? move the cursor close to the point of intersection. Press ENTER .

The word Intersection and the x and y coordinates of the intersection are displayed at the bottom of the screen. In this example you find that the intersection is at x=2 and y=1.

Example 8 Finding a Break-Even Point Graphically

Let's use this example to practice some of the graphing calculator skills you have learned to date. Here is an outline of the process for finding the point of intersection, the break-even point.

1. Identify the data points of the cost function as (3.5, 540) and (1,650).

2. Enter the data points of the cost function into L_1 and L_2.

3. Find the linear regression equation for the cost function.

4. Graph the linear regression equation.

5. Identify the data points of the revenue function as (2.99, 500) and (1.89,600). Enter these into L_1 and L_2.

6. Repeat steps 3 and 4 for the revenue function.

7. Set the window at
    $Xmin=0, Xmax=4, Xscl=1, Ymin=0, Ymax=800, Yscl=100$.

8. Find the point of intersection, the break-even point in this example at $x=1.66, \; y=621$.

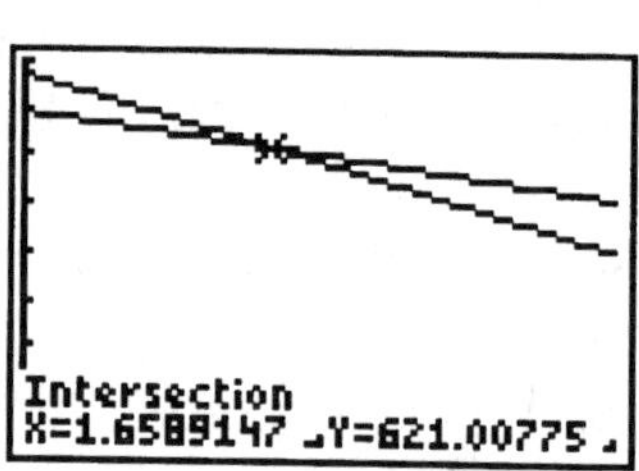

The result is that the break-even point occurs when the price, the x-coordinate, is $1.66 (rounded to the nearest cent) and the access fee, the y-coordinate, is $621 (rounded to the nearest dollar).

2.2 Using Matrices to Solve Systems of Equations
TI-83 Plus

In this section you are shown how to use matrices and their row operations to
solve a system of linear equations. Here let's use 2.2 Example 6 *Row
Operations using Technology* to demonstrate how to do this on the TI-83 Plus.

Example 6: Use technology to solve the system with augmented matrix

        [2/3   0   1   1]
        [3/7   -2/7   1   1]
        [2   1   -5   2]

Enter the Matrix

First you must know how to enter a matrix into the calculator. To do this,
press [2nd] [x⁻¹] (MATRX) key. Notice that there are ten matrix names,
[A],[B],...,[J], available on the TI-83 Plus. A matrix must be named
using one of these names.

Let's use [A] as our first matrix.

To enter the dimension and the elements of the
matrix [A], move the right arrow twice to EDIT.
Select EDIT 1:[A].

Enter the dimension of [A] as 3 x 4 and press
[ENTER] after each entry.

You are now ready to enter the elements of
[A] = [2/3   0   1   1]
      [3/7   -2/7   1   1]
      [2   1   -5   2]

 Notice that at the bottom of the screen you are
given the position of the element as 1,1 or 2,3 etc.
This means that you are in row 1   column 1, or row
2 column 3, and so on. The three dots ... to the right or left of a matrix tell
you that there is more of the matrix to see in that direction.

If appropriate, you may consider changing the MODE to a different Float setting so that you can see fewer decimal places.

Notice that even though you enter the elements in fraction form, the display shows them in decimal form. You may wish to clear the fractions by hand from each row before you enter it.

Let's clear the fractions, and re-enter the matrix [A] with integer values. Also, let's reset the MODE to setting of Float that the word Float is highlighted.

Before we go on, let's view matrix [A] on the Home Screen. Press [2nd] [MODE] (QUIT) to leave the Matrix menu.

You must use [2nd] [x⁻¹] (MATRX) NAMES 1:[A] 3x4 to do this.

Go into the matrix menu and on the NAMES screen select 1:[A] 3x4 . This puts [A] on the Home Screen. Press [ENTER] to view matrix [A].

Row Operations

We are now ready to perform row operations on matrix [A].

Pivot on the 2 in row 1, column 1

Following example 6, we will first pivot around the 2 of row 1, column 1.

$R_2 \rightarrow 2R_2 - 3R_1$

To replace R_2 with $2R_2 - 3R_1$ requires two steps: first find $2R_2$, then add $-3R_1$ to the result.

$2R_2 \rightarrow R_2$

To do this, go to $\boxed{\text{MATRX}}$ MATH. Use the up arrow to see the row operations screen. Select E:*row(. This puts *row(on the Home Screen.

Next, finish the command to read *row(2,[A],2)→[B]. Use the $\boxed{\text{STO►}}$ key to get the → . Store the result into [A] again or another matrix, say [B]. The $\boxed{\text{STO►}}$ key is located above the $\boxed{\text{ON}}$ key and displays an arrow on the screen. Let's use a new matrix name each time.

Don't forget to use $\boxed{\text{2nd}}$ $\boxed{x^{-1}}$ (MATRX) NAMES 1:[A] 3x4 to enter the name of the matrix [A] on the Home Screen. Use $\boxed{\text{2nd}}$ $\boxed{x^{-1}}$ (MATRX) NAMES 2:[B] 3x4 to enter the name of the matrix [B] on the Home Screen.

```
*row(2,[A],2)→[B
]
   [[2  0   3   3 ]
    [6  -4  14  14]
    [2  1   -5  2 ]]
■
```

You do not need to change the dimension of matrix [B]. Using the $\boxed{\text{STO►}}$ key automatically makes matrix [B] a 3x4 matrix since matrix :[A] is 3x4.

$R_2 - 3R_1 \rightarrow R_2$

Now we add $-3R_1$ to the new row 2. Go back to $\boxed{\text{2nd}}$ $\boxed{x^{-1}}$ (MATRX) MATH, select F:*row+(. On the Home Screen finish the command
  F:*row+(-3,[B],1,2) → [C].

```
*row+(-3,[B],1,2
)→[C]
   [[2  0   3   3]
    [0  -4  5   5]
    [2  1   -5  2]]
```

The element in row 2, column 1 is now 0. The new matrix [C] is shown here.

$R_3 - R_1 \rightarrow R_3$

To finish column 1, multiply row 1 by −1, then add it to row 3. Again use F:*row+(. Recall that you can repeat a line using $\boxed{\text{2nd}}$$\boxed{\text{ENTER}}$ (ENTRY). On the Home Screen finish the command
  F:*row+(-3,[C],1,3) → [D].

```
*row+(-1,[C],1,3
)→[D]
   [[2  0   3   3 ]
    [0  -4  5   5 ]
    [0  1   -8  -1]]
■
```

The element in row 3, column 1 is now 0. The new matrix [D] is shown here.

Pivot on the −4 in row 2, column 2

Next, we select the first non-zero number in the second row as the pivot , and clear its column. Hence, pivot on the −4 in column 2. We need to multiply row 3 by 4 and add it to row 2 . However, this must be done in two steps. The following screens show the remaining steps, $4R_3 \rightarrow R_3$ and $R_3 + R_2 \rightarrow R_3$.

$4R_3 \rightarrow R_3$

Multiply row 3 by 4 to get a new row 3. Use
[2nd] [x⁻¹] (MATRX) MATH, select E:*row(.
Finish the command to read
  *row(4,[D],3) →[E].

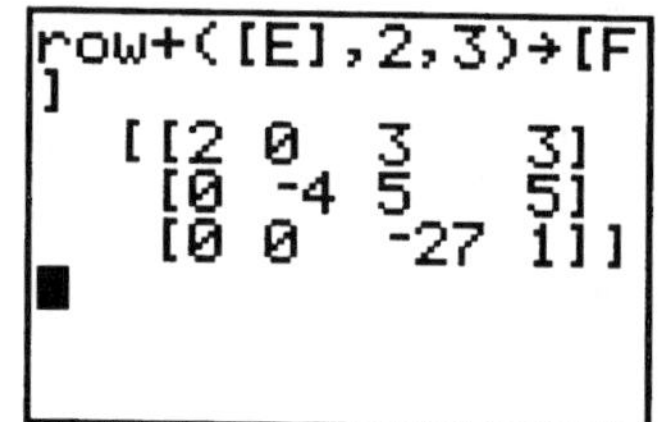

$R_3 + R_2 \rightarrow R_3$

Add rows 2 and 3 to get a new row 3. Use [2nd]
[x⁻¹] (MATRX) MATH, select D:row+(.
Finish the command to read
  row+([E],2,3) →[F].

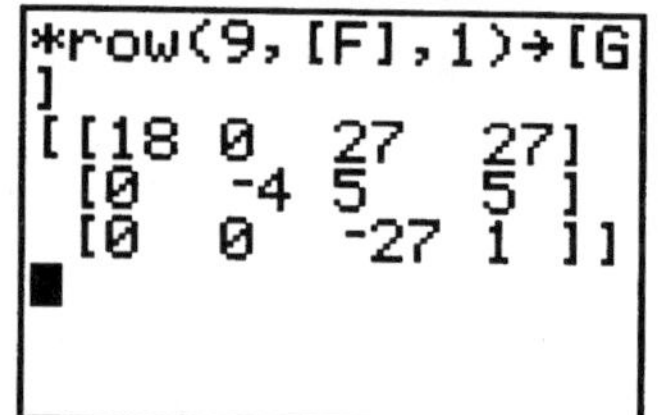

Pivot on the −27 in row 3, column 3

Lastly, we must pivot around the −27 in row 3, column 3. First work with row 1 and change the 3 to a 0. This requires two steps. First, multiply row 1 by 9, then add row 3 to row 1 gives a new row 1, $9R_1 \rightarrow R_1$ and $R_1 + R_3 \rightarrow R_1$.

$9R_1 \rightarrow R_1$

Multiply row 1 by 9 to get a new row 1. Use
[2nd] [x⁻¹] (MATRX) MATH, select E:*row(.
Finish the command to read
  *row(9,[F],1) →[G].

$R_1 + R_3 \rightarrow R_1$

Add rows 1 and 3 to get a new row 1. Use [2nd]
[x⁻¹] (MATRX) MATH, select D:row+(.
Finish the command to read
  row+([G],3,1) →[H].

The last element to change is 5 in row 2, column 3 to 0. This requires three steps: multiply row 2 by −27 to get a new row 2; then multiply row 3 by −5 to get a new row 3. Finally add row 2 and row 3.

$-27R_2 \rightarrow R_2$ and $-5R_3 \rightarrow R_3$ and $R_3+R_2 \rightarrow R_3$.

$-27R_2 \rightarrow R_2$

Multiply row 2 by −27 to get a new row 2. Use [2nd] [x⁻¹] (MATRX) MATH, select E:*row(. Finish the command to read *row(−27,[H],2) →[I].
Use the right arrow to see the

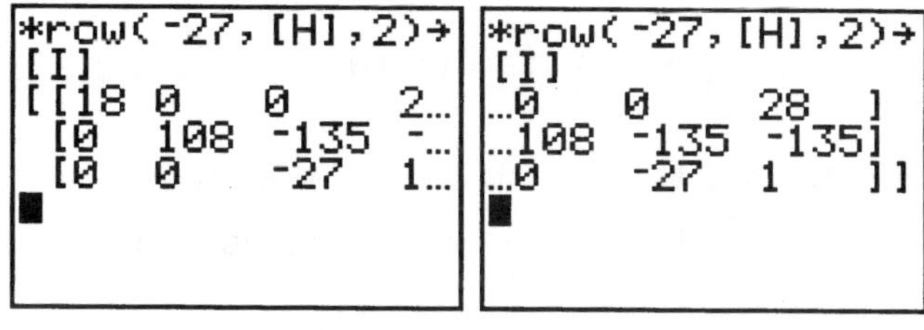

entire matrix. We need two screens here to see the entire matrix.

$-5R_3 \rightarrow R_3$

Multiply row 3by −5 to get a new row 3. Use [2nd] [x⁻¹] (MATRX) MATH, select E:*row(. Finish the command to read *row(−5,[I],3) →[J].
Use the right arrow to see the

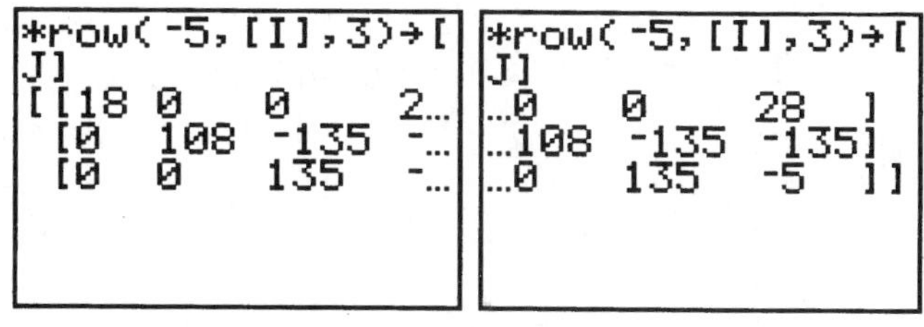

entire matrix. We need two screens here to see the entire matrix.

$R_3 + R_2 \rightarrow R_3$

Add rows 2 and 3 to get a new row 2. Use [2nd] [x⁻¹] (MATRX) MATH, select D:row+(. Finish the command to read row+([J],3,2) →[J].
Use the right arrow to see the

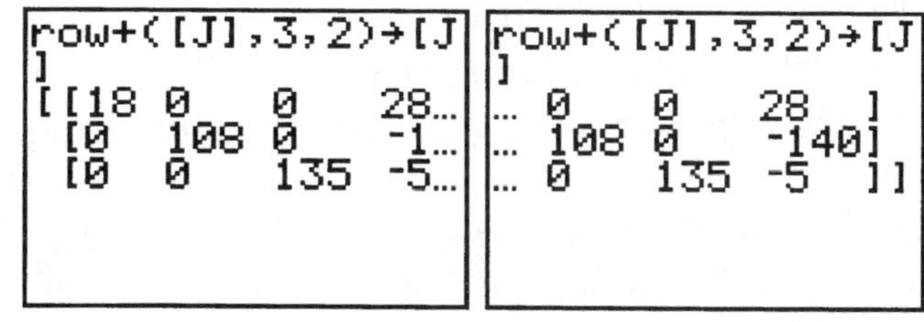

entire matrix. We need two screens here to see the entire matrix.

Finally, read the results as $(x, y, z) = \left(\dfrac{28}{18}, -\dfrac{-140}{108}, \dfrac{-5}{135} \right)$ or

$(x, y, z) = \left(\dfrac{14}{9}, -\dfrac{35}{27}, \dfrac{-1}{27} \right)$ in reduced fraction form.

Summary of Matrices for Section 2.2 on the TI-83 Plus

About matrices on the TI-83 Plus:

1. Matrices must be entered and referred to using the MATRX menu by pressing [2nd] [x⁻¹] (MATRX).
2. Enter a matrix using MATRX EDIT.
3. Refer to a matrix using MATRX NAMES. There are 10 matrix names, [A],[B],...,[J], available on the TI-83 Plus.
4. Perform matrix operations using MATRX MATH.
5. Add, subtract, multiply, and divide matrices on the HOME screen using MATRX NAMES to input each matrix.
6. ... to the right or left of a matrix tells you that there is more of the matrix to see in that direction. If appropriate, you may consider changing the MODE to a different Float setting so that you can see the entire matrix on one screen.
7. Press [2nd] [MODE] (QUIT) to leave the screen. This returns you to the HOME screen.

To enter a matrix on the TI-83 Plus:

1. Press [2nd] [x⁻¹] (MATRX).
2. Cursor to EDIT, press [ENTER]. Select a matrix, say 1:[A] and press [ENTER].
3. Use the arrow keys to change the dimension. Press [ENTER] after each entry.
4. Enter the elements of matrix [A]. Press [ENTER] after each entry.
5. Press [2nd] [MODE] (QUIT) to leave the screen. This returns you to the HOME screen.

To see a matrix, say [A], on the Home Screen:

1. From the HOME screen press [2nd] [x⁻¹] (MATRX).
2. NAMES is highlighted. Select 1:[A] or another matrix . Press [ENTER] to see the name of the matrix on the HOME screen, say [A].
3. Press [ENTER] again to the entries in the matrix.
4. Use [MODE] Float to adjust the number of decimal places you wish to see in the matrix .
5. Use ▸Frac found in [MATH] MATH 1:▸Frac to convert the entries of the matrix to fractions.

Row operations on the TI-83 Plus:

There are four row operations that are used in solving a system of equations by matrices using pivots or the Gauss-Jordan method: rowSwap(, row+(, *row(, and *row+(. I would suggest that you use the [STO▸] key to store the result of each operation as a new matrix. Note that storing a matrix to a new name automatically redefines the dimension of an existing matrix with that name.

rowSwap($R_i \Leftrightarrow R_j$ Swap row i with row j).
rowSwap(matrix name, swap row i, with row j)
Example: rowswap([A],1,2) → [B]   swaps row 1 with row 2

1. Press [2nd] [x⁻¹] (MATRX).
2. Cursor to MATH.
3. Select C:rowSwap, press [ENTER]. This brings you to the HOME screen.
4. Use MATRX NAMES to input the name of the matrix.
5. Enter the rows to be swapped.
6. Use the [STO▸] key and MATRX NAMES to store the result as a new matrix.

row+($R_i + R_j \rightarrow R_j$
Add row i with row j, store the result in row j.
row+ (matrix name, add row i, with row j and store it in row j)
Example: row+ ([B],1,2) → [C] adds rows 1 and 2 of matrix [B] and stores the result in row 2.
1. Press [2nd] [x⁻¹] (MATRX).

2. Cursor to MATH.
3. Select D:row+(, press ENTER. This brings you to the HOME screen.
4. Use MATRIX NAMES to input the matrix.
5. Enter the rows to be added.
6. Use the STO▸ key and MATRX NAMES to store the result as a new matrix.

row(kR_i) k*R_i → R_i

Multiply row i by k, store the result in row i.

***row (constant, matrix name, row i and store it in row i)**

Example: *row (5, [C],1) → [D] multiplies row 1 of matrix [C] by 5 and stores result in row 1.
1. Press 2nd x^{-1} (MATRX).
2. Cursor to MATH.
3. Select E:*row (, press ENTER. This brings you to the HOME screen.
4. Enter the constant multiple.
5. Use MATRX NAMES to input the matrix.
6. Enter the row to be multiplied by the constant.
7. Use the STO▸ key and MATRX NAMES to store the result as a new matrix.

row+(kR_i + R_j → R_j

Multiply row i by k, add the result to row j, store the final result in row j.

***row+ (constant, matrix name, row 1, row 2)**

Example: *row (-3, [D],1,2) → [E]   multiplies row 1 of matrix [D] by -3, add the result to row 2, and store the result in row 2.
1. Press 2nd x^{-1} (MATRX).
2. Cursor to MATH.
3. Select F:*row+ (, press ENTER. This brings you to the HOME screen.
4. Enter the constant multiple.
5. Use MATRX NAMES to input the matrix.
6. Enter the row to be multiplied by the constant.
7. Enter the row to which it is added.
8. Use the STO▸ key and MATRX NAMES to store the result as a new matrix.

ref and rref on the TI-83 Plus

The TI-83 Plus can reduce a matrix to row echelon form using **ref**, and to row reduced echelon form using **rref**. The resulting matrix is not stored back to the original matrix, so it should be stored to a new matrix name.

ref (Matrix Name)
ref([A]) → [B]   returns the row echelon form of matrix [A]

Example: ref([A]) of section 2.2 example 6 is shown in fraction form.
1. Press $\boxed{\text{2nd}}$ $\boxed{x^{-1}}$ (MATRX).
2. Cursor to MATH.
3. Select A:ref (, press $\boxed{\text{ENTER}}$. This brings you to the HOME screen.
4. Use MATRX NAMES to input the matrix.
5. Use the $\boxed{\text{STO▸}}$ key and MATRX NAMES to store the result as a new matrix.

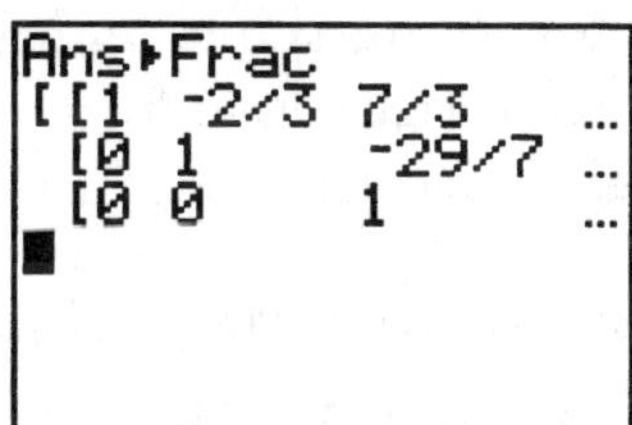
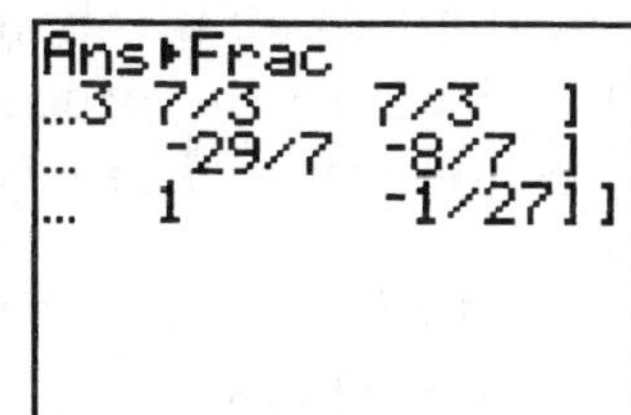

rref(Matrix Name)   (Gauss-Jordan Method)
rref([A]) → [C]   returns the reduced row echelon form of matrix [A]

Example: ref([A]) of section 2.2 example 6 is shown in reduced fraction form.

1. Press $\boxed{\text{2nd}}$ $\boxed{x^{-1}}$ (MATRX).
2. Cursor to MATH.
3. Select B:ref (, press $\boxed{\text{ENTER}}$. This brings you to the HOME screen.
4. Use MATRX NAMES to input the matrix.
5. Use the $\boxed{\text{STO▸}}$ key and MATRIX NAMES to store the result as a new matrix.

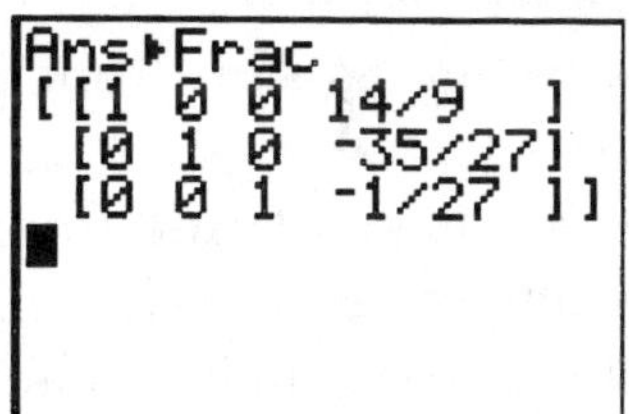

2.3 Applications of Systems of Linear Equations – TI-83 Plus

There are no new calculator skills in this section.

2.4 You're the Expert – The Impact of Regulating Sulfur Emissions – TI-83 Plus

All the skills needed for this section have already been presented. Here is an outline of two methods for using the calculator to solve this problem. You must follow the procedure twice, once for the cost function and again for the jobs function. Let's begin with the cost function.

Method 1: Using Systems of Linear Equations and Matrices

Identify three points of the cost function: $(8, 20.4)$, $(10, 34.5)$, and $(12, 93.6)$ where millions of tons t is the first coordinate and cost C is the second coordinate.

Next, use a quadratic model $C = at^2 + bt + c$ to generate three quadratic equations, one for each point. For example, when t=8 and C=20.4 the quadratic equation $C = at^2 + bt + c$ becomes $20.4 = a(8)^2 + b(8) + c$ or $20.4 = 64a + 8b + c$. Repeat the process for t=10 and t=12.

This gives you three linear equations in three unknowns. Enter the three equations as matrix [C] in the matrix menu. Using [C] as the name of the matrix helps you remember that this is the cost matrix.

Use $\texttt{rref([C])}$ to identify a=5.625, b= -94.5, and c=414 of the quadratic model.

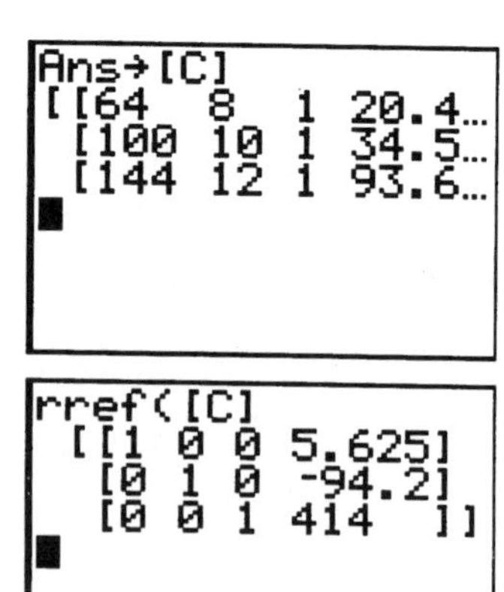

Write the quadratic equation model as $C = 5.625t^2 - 94.2t + 414$ using the results of solving matrix [C].

Lastly, substitute t=15 to the quadratic cost equation to predict the annual cost. To do this, enter `5.625*15`^{2}` - 94.2*15 +414` on the Home screen. Press `ENTER` to see the result 266.625. The annual cost, then, is $266.625 billion.

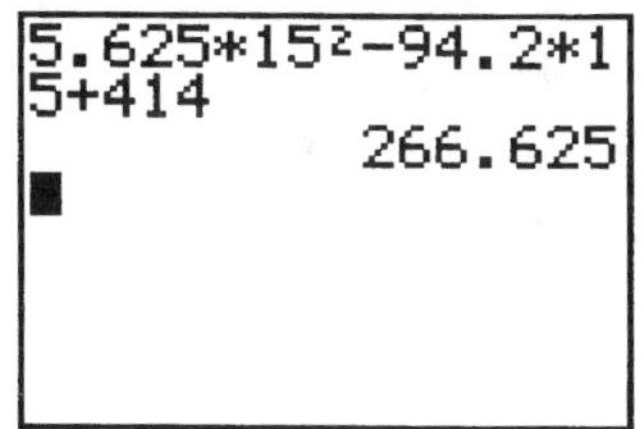

Repeat the entire process for the jobs information.

Identify the jobs points as `(8,14100)`, `(10,21900)`, and `(12,13400)`.

Get the system of three linear equations for the jobs information. Enter the coefficients as a matrix. Reduce the matrix using rref and read off `a=-2037.5`, `b=40575`, and `c=-180100`.

Write the quadratic equation $J = -2{,}037.5t^2 + 40{,}575t - 180{,}100$.
When t=15, jobs J= $-29{,}913$ jobs lost, or a gain of $29{,}913$ jobs.

Method 2: Using the STAT menu

Again we need to do the process twice, once for the cost information, and again for the jobs information.

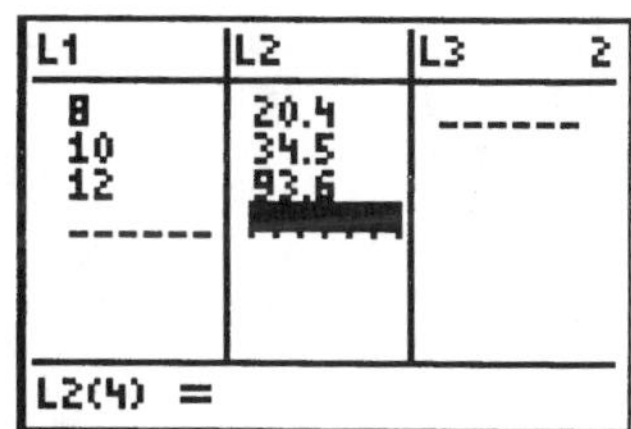

Identify three points of the cost function:
`(8,20.4)`, `(10,34.5)`, and `(12, 93.6)` where millions of tons t is the first coordinate and cost C is the second coordinate. Enter the points into L_1 and L_2 of `STAT` `EDIT`.

Next, find the quadratic regression equation using `STAT` `CALC` `5:QuadReg`. Press `ENTER` to see a, b , and c.

Lastly, substitute t=15 to the quadratic cost equation to predict the annual cost of $266.625. To do this, enter `5.625*15`^{2}` - 94.2*15 +414` on the Home screen as shown above.

Identify the jobs points as $(8, 14100)$, $(10, 21900)$, and $(12, 13400)$. You already have the t-values listed in L_1. Enter the jobs values into L_3.

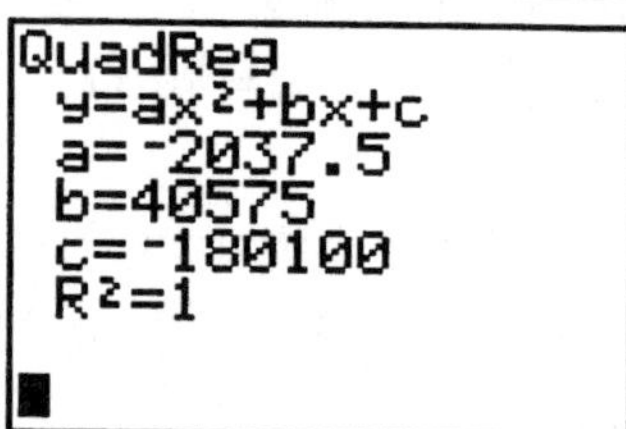

Use [STAT] `CALC 5:QuadReg` as before to put QuadReg on the Home screen. Then input L_1, L_3. To input L_1 and L_3, press [2nd] [1] (L_1), then the comma [,], then [2nd] [3] (L_3).

Press [ENTER] to see the quadratic model for the jobs information.

The quadratic regression equation for L_1 and L_3 jobs information is $J = -2{,}037.5t^2 + 40{,}575t - 180{,}100$.

Evaluate the model on the Home screen to find that when $t = 15$, jobs $J = -29{,}912.5$, an increase of $29{,}913$ jobs.

CHAPTER 3

MATRIX ALGEBRA AND APPLICATIONS

TI-83 PLUS

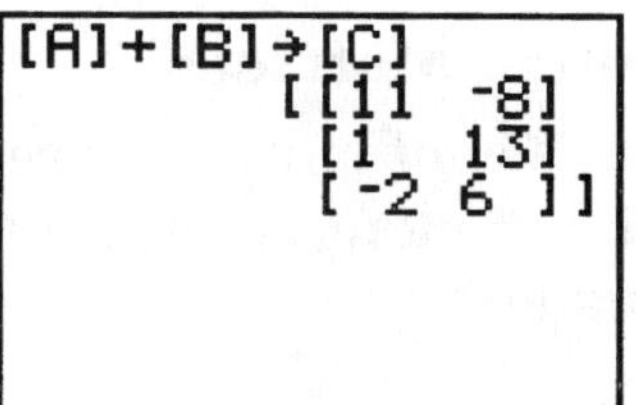

3.1 Matrix Addition and Scalar Multiplication – TI-83 Plus

In Chapter 2 you saw how to name a matrix and input its dimensions and entries by using MATRX EDIT. You also saw how to use a matrix on the Home screen using MATRX NAMES. See the *Summary of Matrices for Section 2.2 on the TI-83 Plus* at the end of section 2.2.

In this section you will see how to decide whether or not two matrices are equal, add and subtract matrices of the same size (dimensions), perform scalar multiplication i.e. multiply a matrix by a constant value k, and find the transpose of a matrix, i.e. interchange the rows and columns.

Test the Equality of Matrices using the Test menu

The equality of matrices may be tested using the TEST menu. A result of 1 indicates a YES or TRUE answer to the comparison, a result of 0 indicates a NO or FALSE answer to the comparison. Only the = and ≠ selections on the TEST menu can be used with matrices. The two matrices must have the same dimension. If they are not the same dimension, you will get an ERR: DIM MISMATCH message.

To do this, input or select a matrix, say [A], on your calculator using MATRX NAMES. Next, press 2nd MATH (TEST)1:= ENTER. Next input or select a second matrix, say [B]. Press ENTER to see 0, 1, or ERR: DIM MISMATCH. Answers will vary according to the matrices chosen.

Addition and Subtraction of Matrices

Matrix addition and subtraction of matrices are performed on the HOME screen using MATRIX NAMES to identify the matrices. Ordinary arithmetic operations are entered as usual.

Let's use the two Quick Examples given in *Matrix Addition and Subtraction*. Enter the first matrix as [A] and the second matrix as [B]; store the result as [C].

```
[A]+[B]→[C]
          [[11  -8]
           [1   13]
           [-2  6 ]]
```

$$1. \quad \begin{bmatrix} 2 & -3 \\ 1 & 0 \\ -1 & 3 \end{bmatrix} + \begin{bmatrix} 9 & -5 \\ 0 & 13 \\ -1 & 3 \end{bmatrix} = \begin{bmatrix} 11 & -8 \\ 1 & 13 \\ -2 & 6 \end{bmatrix}$$

To add these on the Home screen, use MATRX NAMES input [A]. Press +, then use MATRX NAMES input [B]. Press STO▶ then MATRX NAMES input [C].

$$ 2. \quad \begin{bmatrix} 2 & -3 \\ 1 & 0 \\ -1 & 3 \end{bmatrix} - \begin{bmatrix} 9 & -5 \\ 0 & 13 \\ -1 & 3 \end{bmatrix} = \begin{bmatrix} -7 & 2 \\ 1 & -13 \\ 0 & 0 \end{bmatrix} $$

To subtract these on the Home screen, use MATRX NAMES input [A].

Press −, then use MATRX NAMES to input [B].

Press STO▶ then MATRX NAMES to input [D].

Since the same matrices are used in both examples, a quicker way to do this is to repeat the line by pressing 2nd ENTER (ENTRY). Next, use the left arrow key ◀ to position the cursor over the plus sign. Press the subtraction key −. Then use the right arrow key ▶ to position the cursor over the [C]. Use MATRX NAMES to input [D]. Press ENTER to see the result.

Scalar Multiplication of Matrices

Scalar multiplication is also performed on the Home screen using MATRX NAMES to identify the matrices. A scalar is an individual number that multiplies a matrix. Let's use matrix [A] from the addition and subtraction example to illustrate scalar multiplication to evaluate 6[A].

To evaluate 6[A] first enter the 6 then use MATRX NAMES to input [A]. Press ENTER to see the result. Notice that it not necessary to use the multiplication key, ×, to do scalar multiplication.

Find the Transpose of a Matrix

The transpose of a matrix is found by writing the rows of the matrix as columns.

The transpose operation is found in MATRX MATH 2:^{T}.

Let's find the transpose of Quick Example 1, where

$$B = \begin{bmatrix} 2 & 3 \\ 10 & 44 \\ -1 & 3 \\ 8 & 3 \end{bmatrix}. \quad \text{Then } B\text{:}^T = \begin{bmatrix} 2 & 10 & -1 & 8 \\ 3 & 44 & 3 & 3 \end{bmatrix}.$$

To do this on the calculator, input matrix [B].

On the Home screen use MATRX NAMES to input [B].

Next, use MATRX MATH 2:^{T} to put the transpose symbol on the Home screen.

Press [ENTER] to see the result.

3.2 Matrix Multiplication – TI-83 Plus

Matrices may be multiplied provided that the number of *rows of the matrix on the left* is the same as the number of *columns of the matrix on the right*. If this is not the case, then you get a dimension mismatch error message `ERR:DIM MISMATCH`. Once the matrices are entered, you can multiply them on the Home screen using `MATRIX NAMES` either as `[A][B]` or as `[A]*[B]`.

```
ERR:DIM MISMATCH
1:Quit
2:Goto
```

Let's begin by using the two matrices `[A]` and `[B]` that we used to demonstrate addition and subtraction in 3.1.

$$[A] = \begin{bmatrix} 2 & -3 \\ 1 & 0 \\ -1 & 3 \end{bmatrix} \qquad [B] = \begin{bmatrix} 9 & -5 \\ 0 & 13 \\ -1 & 3 \end{bmatrix}$$

When you input `[A][B]` or `[A]*[B]` you get the error message `ERR:DIM MISMATCH`.

Now let's do Example 3(a) *Matrix Product*. Enter the two matrices as `[A]` and `[B]`.

$$[A] = \begin{bmatrix} 2 & 0 & -1 & 3 \\ 1 & -1 & 2 & -2 \end{bmatrix}, \quad [B] = \begin{bmatrix} 1 & 1 & -8 \\ 1 & 0 & 0 \\ 0 & 5 & 2 \\ -2 & 8 & -1 \end{bmatrix}$$

Matrix [A] is the matrix at the top of the screen. On the Home screen enter `[A][B]` or as `[A]*[B]` to see the product as shown.

```
  [[2  0   -1   3 ]
   [1  -1  2   -2]]
[B]
   [[1    1  -8]
    [1    0   0 ]
    [0    5   2 ]
    [-2   8  -1]]
■
```

```
[A][B]
   [[-4  21  -21]
    [4   -5  -2 ]]
```

The Identity Matrix

The square identity matrix of a given dimension, 2x2, 3x3, 4x4, etc. can be entered on the calculator using MATRX MATH 5:identity(. This puts identity(on the Home screen.

Input the size of the matrix as a single number. For example, to see a 4x4 identity matrix, enter [2nd] [x⁻¹] (MATRX)MATH. Select 5:identity(. Press [4] to get identity(4 on the Home screen. Press [ENTER] to view the matrix. Store the identity matrix to a matrix name if you will be using it in a matrix operation.

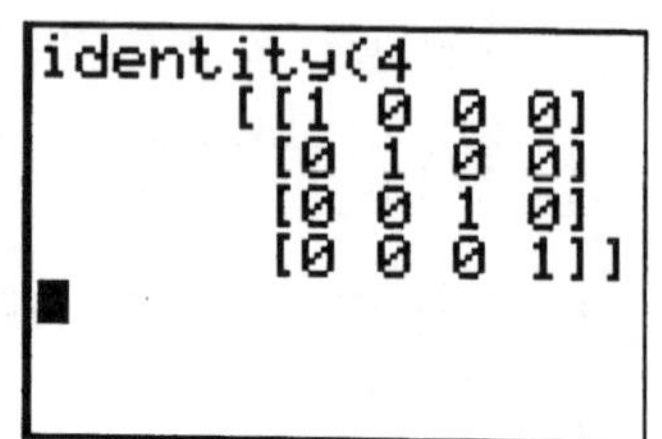

3.3 Matrix Inversion – TI-83 Plus

The process of finding the inverse of a square matrix can be done in two
ways. One way is to use the $\boxed{x^{-1}}$ key. Another way is to augment the identity
matrix of that size to the original matrix using `MATRX MATH 7:augment`
and then perform row operations or `rref`.

Find the inverse of a square matrix using $\boxed{x^{-1}}$

Let's use Example 2 (a) *Computing the Inverse*.
First enter matrix `[A]`. On the Home Screen use
`MATRX NAMES` to input `[A]` then press $\boxed{x^{-1}}$ $\boxed{\text{ENTER}}$ to
see the inverse of `[A]`, denoted by `[A]`$^{-1}$. Store
the result as a new matrix, say `[C]`.

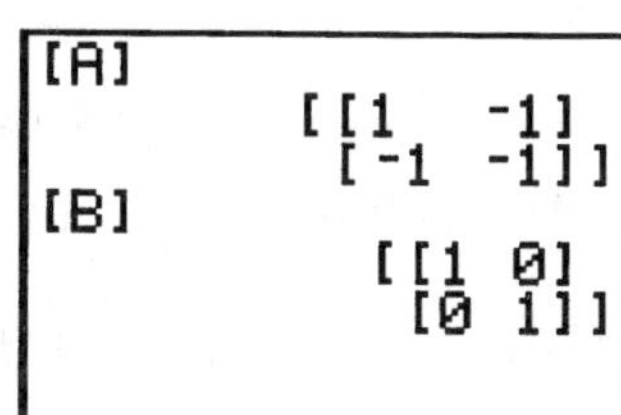

Find the Matrix inverse using `MATRX MATH 7:augment` and `rref`.

Let's again use Example 2 (a) *Computing the
Inverse*. First enter matrix `[A]`.

Next find the 2x2 identity matrix using `MATRX`
`MATH 5: identity (2` as described in section
3.1 and store it as matrix `[B]`.

Now select `MATRIX MATH 7:augment(` to put
`augment(` on the Home screen.

Use `MATRX NAMES` to input `[A]`, `[B]`. Press
$\boxed{\text{ENTER}}$ to see the new matrix. Store the augmented
matrix as `[C]`.   You can also do this on one line
as shown.

Finally, use rref `([C]` or the row methods of
Chapter 2 to get the inverse matrix of [A]. Store
the result as matrix `[D]`.

Let's practice with Example 4 *Solving a Linear System Using the Inverse.*
Here are the steps:

Enter matrices [A] and [B] using MATRX EDIT as shown.

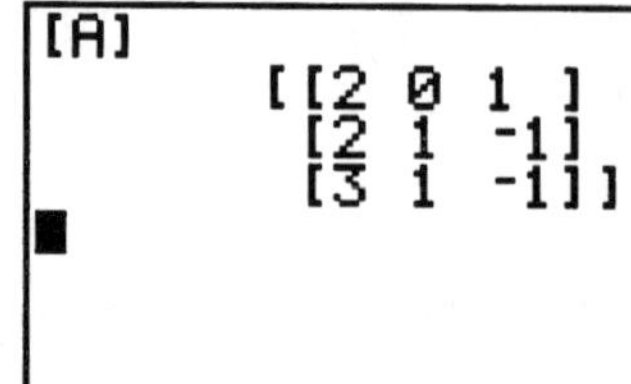
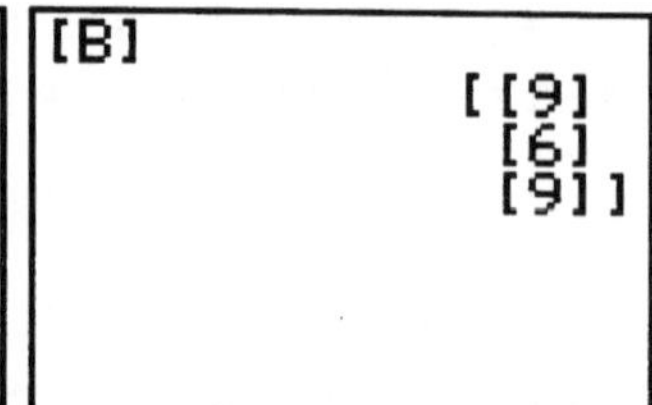

On the Home screen find [A]⁻¹ . Use 2nd x⁻¹ (MATRX) NAMES to put [A] on the Home screen. Use the x⁻¹ to find its inverse. Press ENTER to see the result as shown.

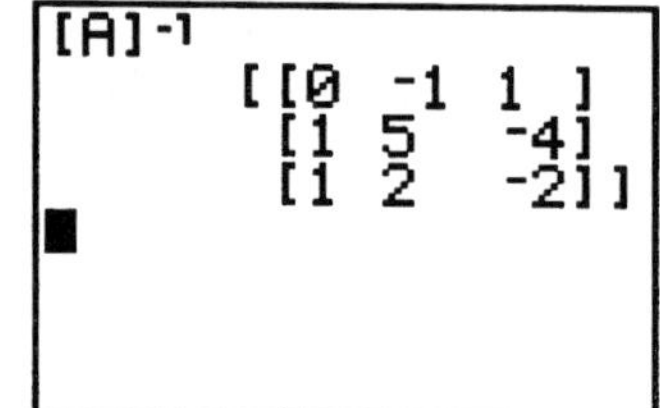

On the Home screen multiply matrices [A]⁻¹ and [B] as shown.

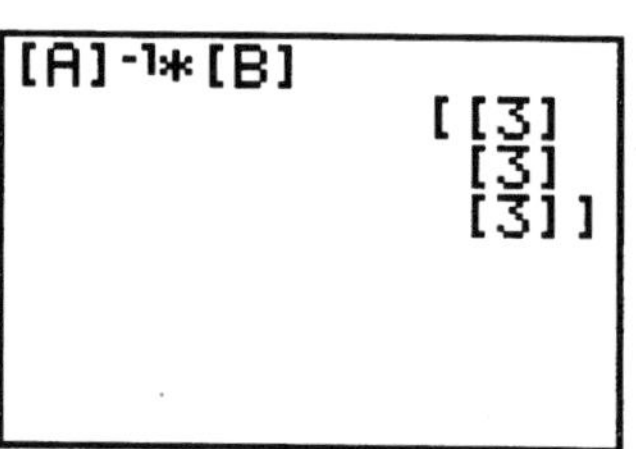

Use 2nd x⁻¹ (MATRX) NAMES to put [A] on the Home screen.

Use the x⁻¹ to find its inverse. Multiply by [B] using 2nd x⁻¹ (MATRX) NAMES. Press ENTER to see the result as shown.

3.4 Input-Output Models – TI-83 Plus

There are no new calculator skills presented in this section. However, let's outline how solve Example 3 *Kenya Economy.*

First use MATRX EDIT to enter matrix [A], the quotients (column entry/column total).   Here is matrix [A] shown over two screens. Notice that you can scroll through the matrix using the left and right arrow keys.

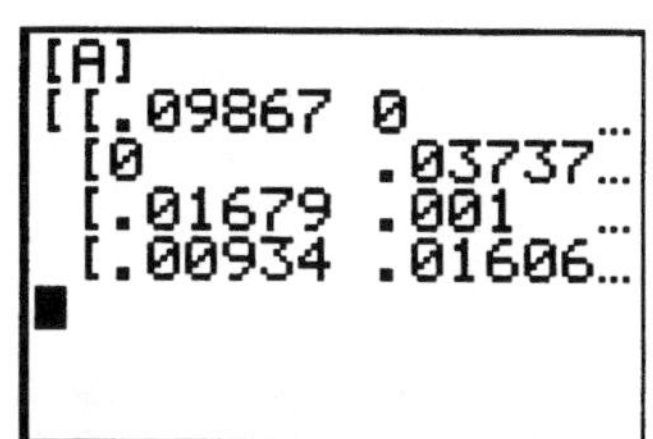

Then enter matrix D^+ as [D].

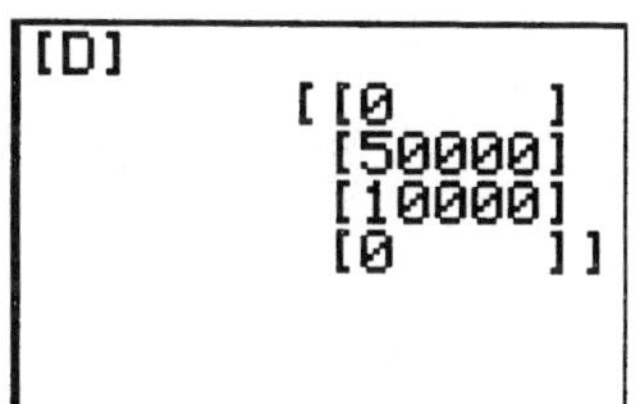

Next you can enter the formula $(I-A)^{-1}D^+$ on one line as shown on the screen.

Round the results to the nearest integer or reset Float to Float 0 and press ENTER.

You're the Expert – The Japanese Economy

There are no new calculator skills in this section.

CHAPTER 4

LINEAR PROGRAMMING

TI-83 PLUS

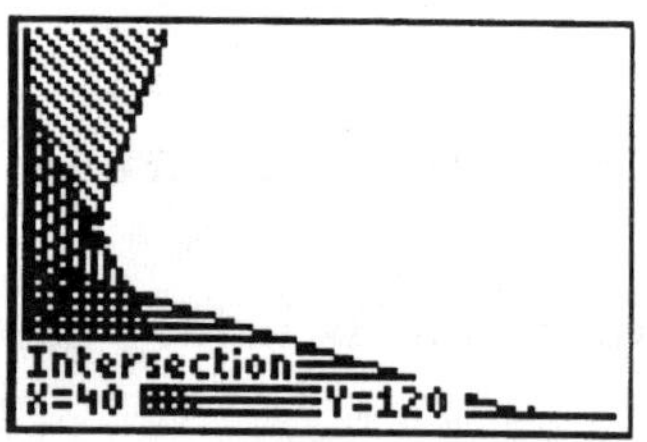

4.1 Graphing Linear Inequalities – TI-83 Plus

Graphing linear inequalities on the TI-83 Plus can be done rather easily.
Most of the work, namely solving for y, and using a test point to determine
which side of the graph must be shaded, is done by you, by hand. Once you
have done this, input the equation and set the icon to shade the solution
area. Here are the steps for the first example presented in this section, solve
the linear inequality $2x + 3y \leq 6$.

To solve $2x + 3y \leq 6$ with the graphing
calculator you must first solve the corresponding
equation, $2x + 3y = 6$ for y.

Then enter the resulting function as
$Y_1 = -(2/3)X + 2$ in the Y= menu.

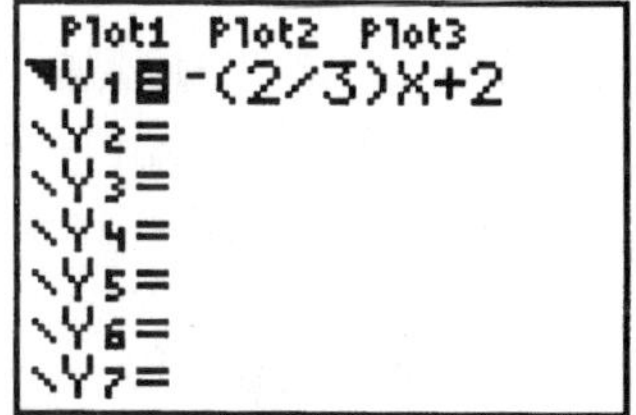

Be sure to use the negation key [(-)] to input the negative sign. Also, notice
that the fraction $-2/3$ is enclosed in parentheses. This is good practice but
not required for the TI-83 Plus. Be aware, though, that on some calculators
this *is* absolutely necessary for the calculator to know that $2/3$ is the
coefficient of x. Otherwise, it thinks that -2 is the numerator and $3x$ is the
denominator, producing a rational function.

 Also notice the = sign is dark. This *is* absolutely necessary to see the graph
on the TI-83 Plus. If the = sign is not dark, move the cursor over the = sign
and press [ENTER].

Next, to tell the calculator to shade the region above of the function, move the
cursor over the ` and press [ENTER] until you see the icon ◥.

You may wish to use the [ZOOM] ZOOM 6:Zstandard window to view the
graph of this function and the shading. Press
[GRAPH] to see the graph and its shading as shown.

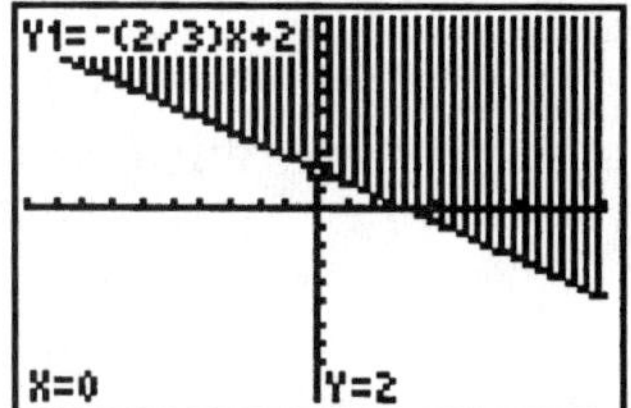

Press [TRACE] to view the equation and the
coordinates of the cursor.

If you do not see the equation and you wish to do so, press [2nd] [ZOOM] (FORMAT), move the cursor to ExprOn and press [ENTER].

Press [GRAPH] to return to the graph and press [TRACE] to see the equation.

Vertical Lines from the Graph screen

Vertical lines are not functions, so they cannot be graphed or shaded on the calculator by using [Y=]. Vertical lines, however, can be *drawn* on the graph using the DRAW menu, found by pressing [2nd] [PRGM] (DRAW).

To do this, select DRAW 4:Vertical and press [ENTER] . Move the vertical line across the screen with the right and left arrow keys.

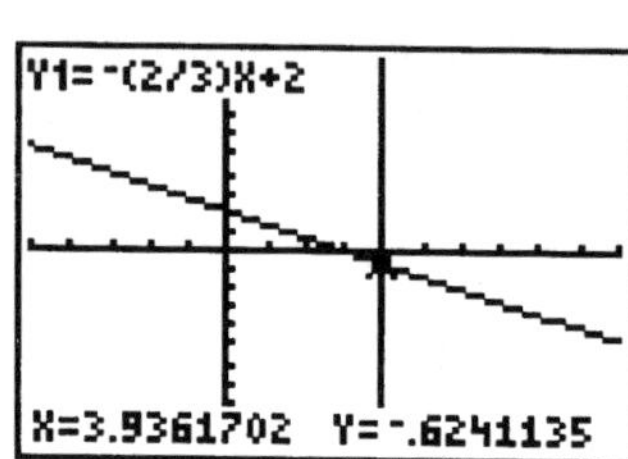

Drawings are purely visuals, therefore no calculations can be done on a drawing. Press [ENTER] to lock in the vertical line. Once the line is locked in, you can use [TRACE] to estimate what you want, say an intersection.

To clear the drawing you must use [2nd] [PRGM] (DRAW) DRAW 1:ClrDraw.

Vertical Lines from the Home screen

You can also draw a stationary (non-movable) vertical line from the Home screen at a specific value of x.

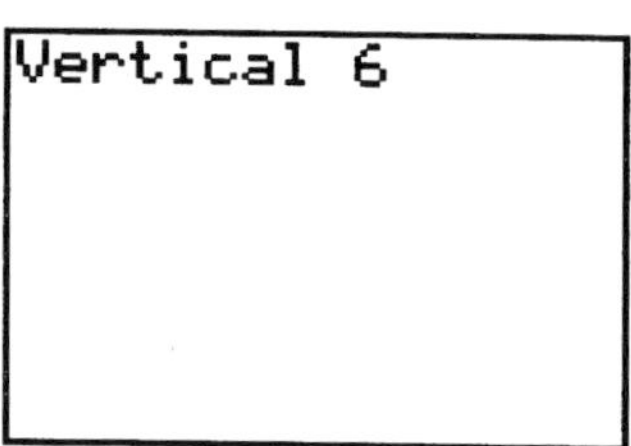

To do this, press [2nd] [PRGM] (DRAW). Select DRAW 4:Vertical. This puts the word Vertical on the Home screen. Enter a single value, say 6, to draw the line x=6. Press [ENTER] to see the graph of x=6.

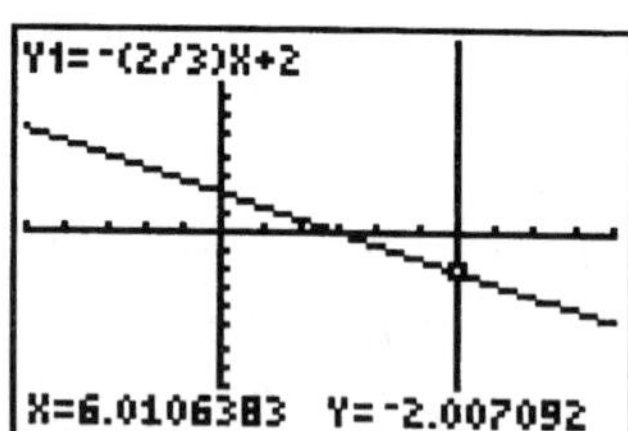

To clear the drawing you must use [2nd] [PRGM] (DRAW) DRAW 1:ClrDraw.

Example 3 *Corner Points*

Example 3 asks you to sketch the region of solutions of the following system
of inequalities, and to list the coordinates of all the corner points.

$$3x - 2y \le 6$$
$$x + y \ge -5$$
$$y \le 4$$

To do Example 3 on the graphing calculator you must first solve each of the
inequalities for y and then graph the
corresponding equations with shading.

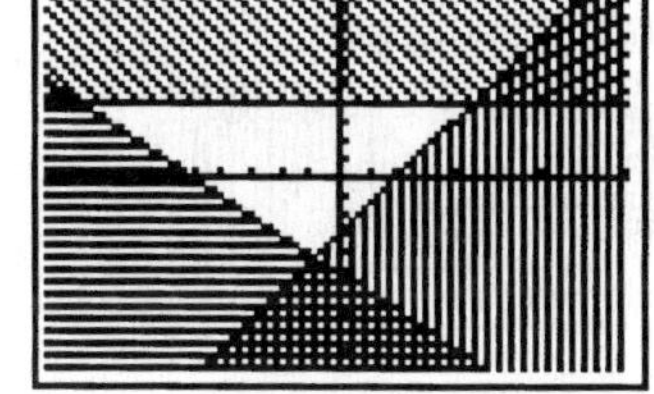

To decide on the shading, use a test point, then
shade the region that is *not* in the solution of the
linear inequality. Recall that the "white" area *is*
the solution of the system.

Input $\quad 3x - 2y \le 6 \quad$ as $\quad$ ▙ $Y_1 = (3/2)x - 3$,

$\qquad\qquad x + y \ge -5 \quad$ as $\quad$ ▙ $Y_2 = -x - 5$

$\qquad\qquad y \le 4 \qquad\quad$ as $\quad$ ▛ $Y_3 = 4$.

Use the standard window by pressing ZOOM 6 .

To set the shading you want, go back to the Y= screen. For each function,
move the cursor over the icon to the left of the $\backslash$Y and press ENTER until you
see the icon you want.

The ▙ icon shades the region below the line, the ▛ shades the region above
the line. Remember that you are shading the region that is *not* the solution
to the linear inequality.

Next, find the corner points by using 2nd TRACE (CALC) 5:intersect
ENTER .

You are then asked to choose the two curves of the intersection or corner you
want. Pay attention to the equation of the curve displayed at the top of the
screen each time you are asked a question such as First curve?, Second
curve? If you need to change the curve, use the up or down arrow until you
see the one you want. Press ENTER after each selection.

When you are asked for a Guess?, move the cursor close the intersection you want, and press [ENTER].

Repeat the process for each corner. Record the results.

Here are the three screens you will see. Can you determine which equations are used for each screen?

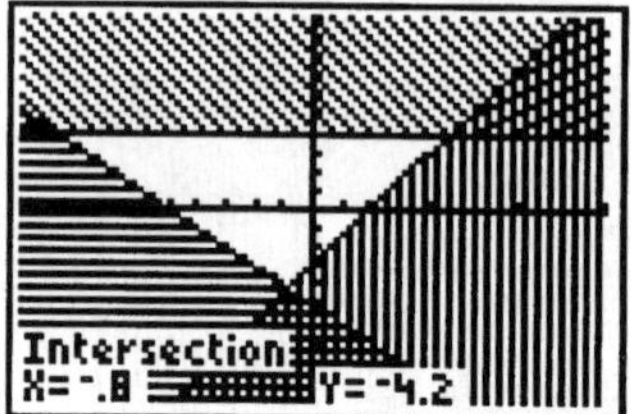
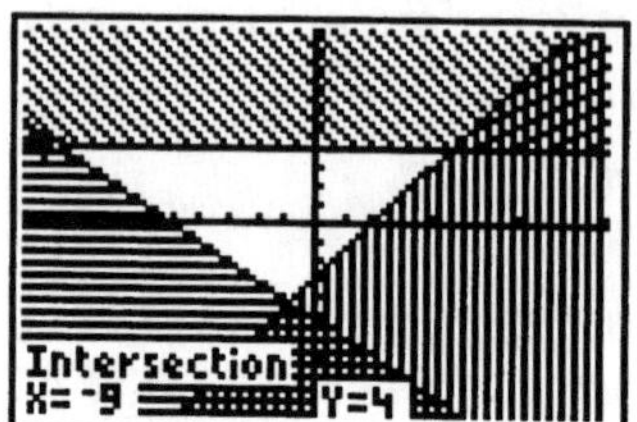
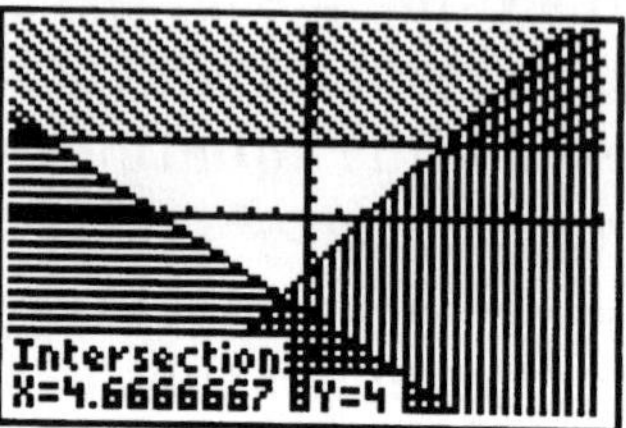

4.2 Solving Linear Programming Problems Graphically – TI-83 Plus

Example 6 *Using the TI-83 Plus to Solve a Linear Programming Problem*

Example 6 uses Excel to minimize the Resource Allocation linear programming problem with the following constraints.

$$\text{Minimize } c = 200x + 400y \qquad \text{Objective function}$$
$$\text{Subject to} \quad 2x + y \geq 200 \qquad \text{Constraint 1}$$
$$x + 3y \geq 300 \qquad \text{Constraint 2}$$
$$3x - y \geq 0 \qquad \text{Constraint 3}$$
$$x \geq 0, \ y \geq 0 \qquad \text{Constraints 4 and 5}$$

The TI-83 Plus does not have a solver feature as described in this example in the text. However, it is possible to solve this problem by graphing the inequalities as described earlier and finding the corners or intersections.

Once the corners are identified, evaluate each corner in the objective function (the equation to minimize) and choose the set of points that gives the smallest (minimum value). Here is an outline of the steps to follow:

Solve each inequality for y. Enter the three inequalities as shown on the screen.

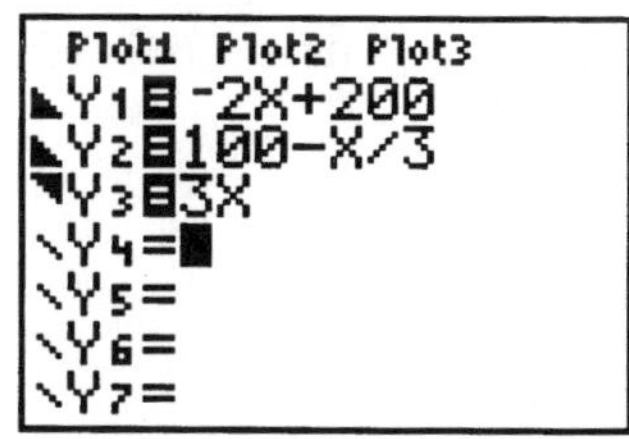

Set the window to Xmin=0, Xmax=350, Xscl=100, Ymin=0, Ymax=250, Yscl=100, Xres=1.

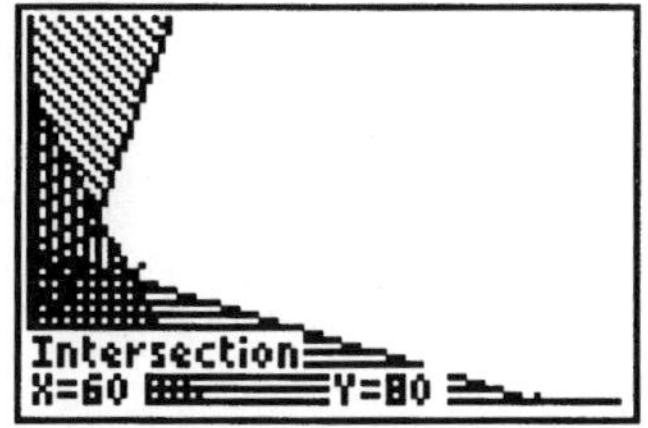

Press [GRAPH] to display the graphs. Note that the last two constraints, namely, x ≥ 0 and y ≥ 0 are used when we set Xmin=0 and Ymin=0.

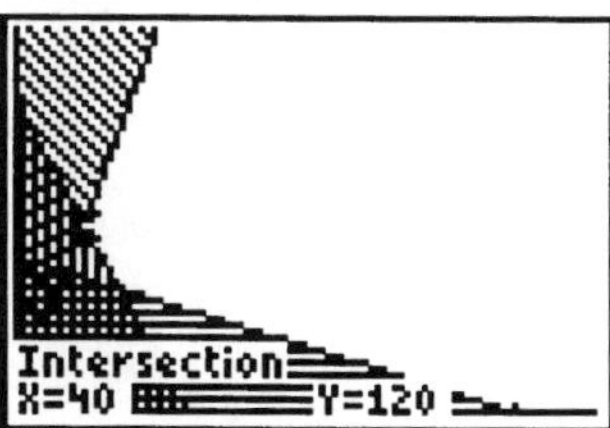

There are three corners (intersections) to identify. Two are intersections of $Y=$ lines, the third is the x-intercept of Y_2.

To find the intersections use [2nd] [TRACE] (CALC) 5:intersect two times as shown on the following screens. Record the results each time.

The third intersection occurs when $y=0$ on Y_2, so we are really looking for an x-intercept, called a zero on the TI-83 Plus. To find the last corner, then, use [2nd] [TRACE] (CALC) 2:zero. Notice that there are three zeros or (x-intercepts). You want the one that is the farthest to the right on the screen on $Y_2 = 100 - x/3$. Be sure to identify the left bound, then, as somewhere close to the left of that point.

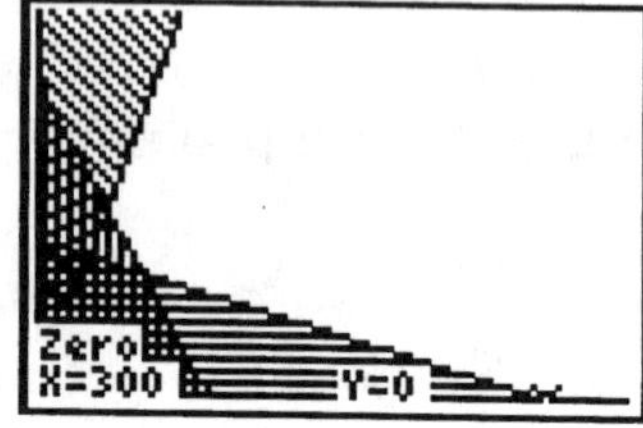

Now that we have identified the corner points as (40,120), (60,800), and (300,0) we must substitute each set of points into the objective function and select the pair that gives the smallest result. You can do this on the Home screen one pair at a time, or use braces and a list of x's and y's as shown. You must scroll through the

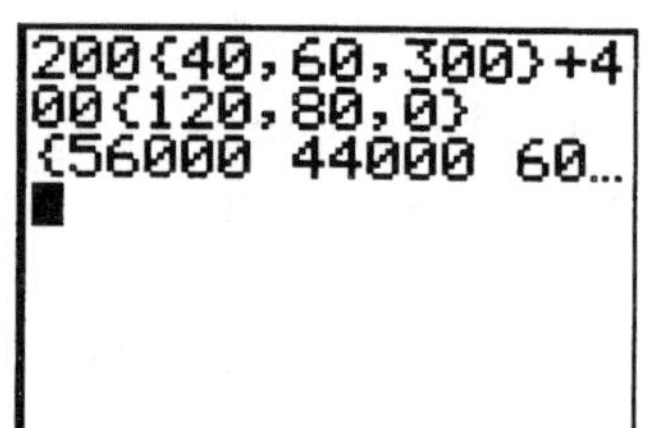

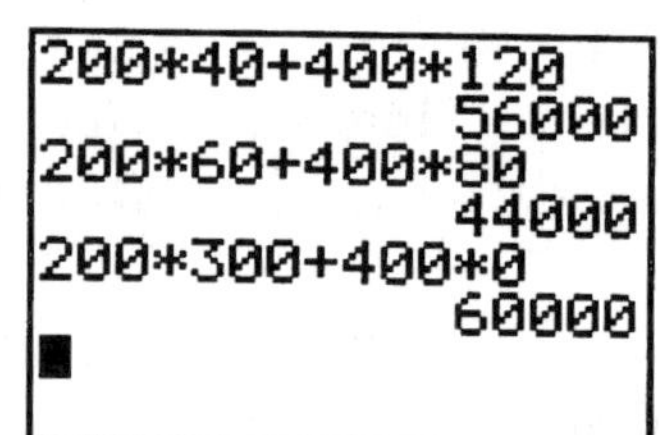

results, though, to see them all. Select the smallest number, 44,000, found using the point (60,80).

4.3 The Simplex Method: Solving Standard Maximization Problems – TI-83 Plus

The Simplex Method returns to the technique of pivoting that was presented in Chapter 2. Recall that the pivot element is used to make all the elements of the pivot column into 0's. Before we look at how to do the Simplex Method on the graphing calculator, let's review the row operations of matrices presented in Chapter 2.

Row operations of matrices on the TI-83 Plus:

All row operations of matrices are found using $\boxed{\text{2nd}}$ $\boxed{x^{-1}}$ (MATRX) MATH. Use the up arrow to locate the command you want. Press $\boxed{\text{ENTER}}$. This brings you to the HOME screen. Follow the formats summarized below using MATRX NAMES when you wish to input the name of a matrix. Use the $\boxed{\text{STO►}}$ key and MATRX NAMES to store the result as a new matrix. Note that ref(and rref(are not useful for the Simplex Method because the pivot column is not always the first column.

rowSwap($R_i \Leftrightarrow R_j$ Swap row i with row j).
rowSwap(matrix name, swap row i, with row j)
Example: rowswap([A],1,2) $\rightarrow$ [B]

        Swap row 1 with row 2

row+($R_i + R_j \rightarrow R_j$
Add row i with row j, store the result in row j.
row+ (matrix name, add row i, with row j and store it in row j)
Example: row+ ([B],1,2) $\rightarrow$ [C]

        Add rows 1 and 2 of matrix [B] and store the result in row 2.

***row(k*R_i) k*$R_i \rightarrow R_i$**
Multiply row i by k, store the result in row i.
***row (constant, matrix name, row i and store it in row i)**
Example: *row (5, [C],1) $\rightarrow$ [D]

        Multiply row 1 of matrix [C] by 5 and store the result in row 1.

***row+(k*R$_i$ + R$_j$ → R$_j$**
Multiply row i by k, add the result to row j, store the result in row j.
***row+ (constant, matrix name, row 1, row 2)**
Example: *row (-3, [D],1,2) → [E]
       Multiply row 1 of matrix [D] by -3,
          add the result to row 2, and
          store the result in row 2.

Example 4 *Using Pivoting Technology*

Maximize $p = 1.3x + 4.5y + 11.3z$ subject to

$$2.4x - 0.21y + 11.7z \le 3$$
$$3.3x + 5.6y - 2.35z \le 5$$
$$x - y + 2.3z \le 10$$
$$x \ge 0,\ y \ge 0,\ z \ge 0$$

To do this example on the graphing calculator we will use our knowledge of matrices and row operations of matrices.

First enter the tableau as matrix [A].   To do this, [2nd] [x^{-1}] (MATRX) EDIT [A] [ENTER].   Enter the dimension as 4 x 8, then enter the elements of the tableau. When you are finished, press [2nd] [MODE] (QUIT) to return to the Home screen.

$$
\begin{bmatrix}
2.4 & -0.21 & 11.7 & 1 & 0 & 0 & 0 & 3 \\
3.3 & 5.6 & -2.35 & 0 & 1 & 0 & 0 & 5 \\
1 & -1 & 2.3 & 0 & 0 & 1 & 0 & 10 \\
-1.3 & -4.5 & -11.3 & 0 & 0 & 0 & 1 & 0
\end{bmatrix}
$$

When you finish entering the tableau as matrix [A], check for accuracy by viewing [A] on the Home screen. To do this press [2nd] [x^{-1}] (MATRX) NAMES 1:[A][ENTER]. Use the left and right arrow keys to scroll through the matrix. Here is how matrix [A] looks on the calculator over three screens.

```
[A]
[[2.4     -.21  11.…
 [3.3     5.6   -2.…
 [1       -1    -2.…
 [-1.3    -4.5  -11…
```

```
[A]
…11.7    1  0  0  0 …
…-2.35   0  1  0  0 …
…-2.3    0  0  1  0 …
…-11.3   0  0  0  1 …
```

```
[A]
…7    1  0  0  0  3 ]
…35   0  1  0  0  5 ]
…3    0  0  1  0  10]
….3   0  0  0  1  0 ]]
```

Next, identify the pivot element as 11.7 in position 1,3 (row most negative entry in the 1, column 3) because the bottom row is −11.3, and the smallest test ratio is 3/11.7 or 0.2564.

Convert the pivot element to 1 in position 1,3 and then use it to change the remaining entries of the pivot column to 0. To do this, press 2nd x⁻¹ (MATRX), cursor over to MATH, then scroll to find E*row(. Press ENTER to put *row(on the Home screen. Then, multiply row 1 by 1/11.7 to convert the pivot element to a 1 using
$$*row(1/11.7,[A],1)\rightarrow[B].$$

The new tableau looks like this.

```
[[0.20513   -0.01795   1       0.08547  0   0   0   0.25641]
 [3.3        5.6      -2.35    0        1   0   0   5      ]
 [1         -1         2.3     0        0   1   0   10     ]
 [-1.3      -4.5     -11.3     0        0   0   1   0      ]]
```

When viewing matrix [B] on the calculator I would suggest recording one column at a time and then check for accuracy again when you are done. The screens would look something like this. Also, if you have not already done so, I would suggest that you set Float to the word Float and round as necessary on your paper.

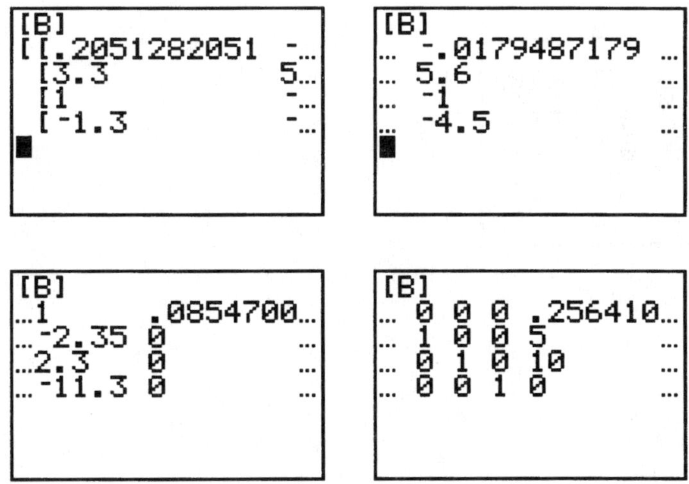

Next, clear column 3, i.e. make the remaining elements of column 3 into 0's using the following commands.

*row+(2.35,[B],1,2)→[C] Multiply row 1 by 2.35, add to row 2 → row 2

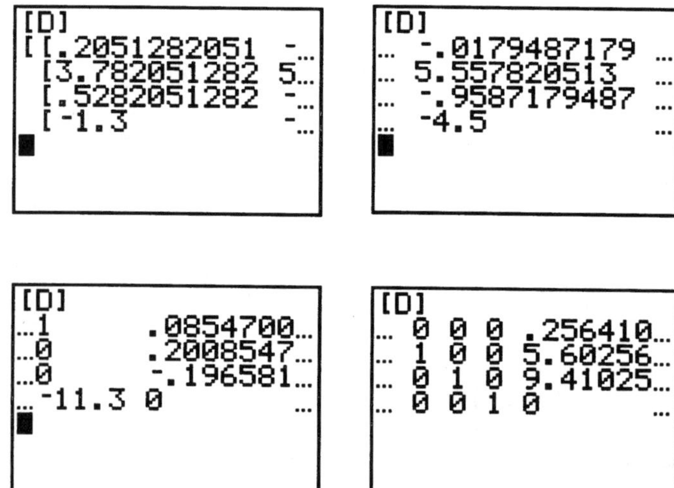

*row+(-2.3,[C],1,3)→[D] Multiply row 1 by −2.3, add to row 3 → row 3 `*row+(11.3,[D],1,4)→[E]` Multiply row 1 by 11.3, add to row 4 → row 4

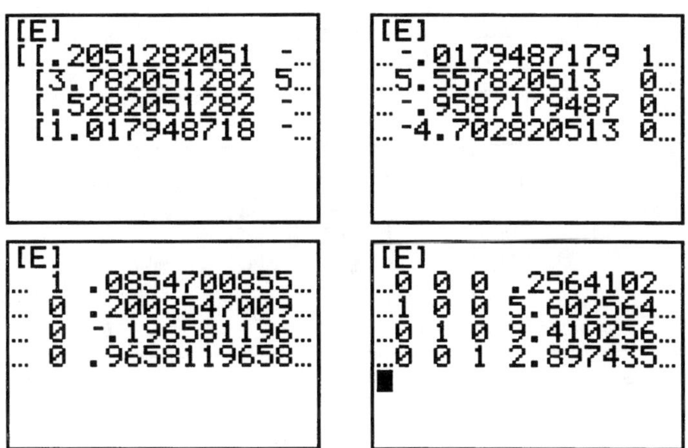

Repeat the process, pivoting around 5.55782 in position 2,2 of the matrix, namely row 2, column 2 using the following commands.

Reset `MODE` to `Float  6`
`*row(1/5.55782,[E],2)→[F]`

*row+(0.017948,[F],2,1)→[G]

These screens are getting rather hard to read. For the remaining work, let's change the MODE to Float 2.

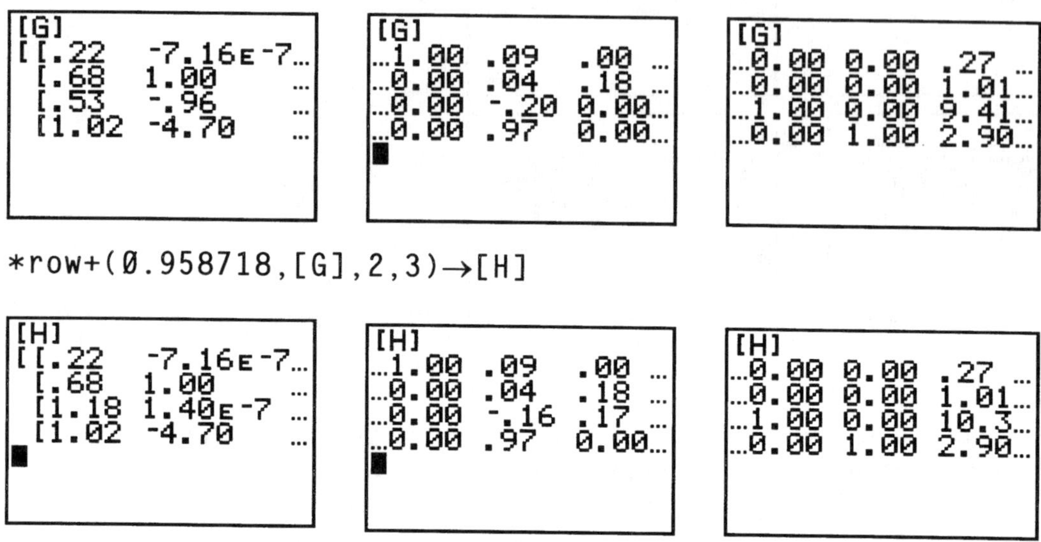

*row+(0.958718,[G],2,3)→[H]

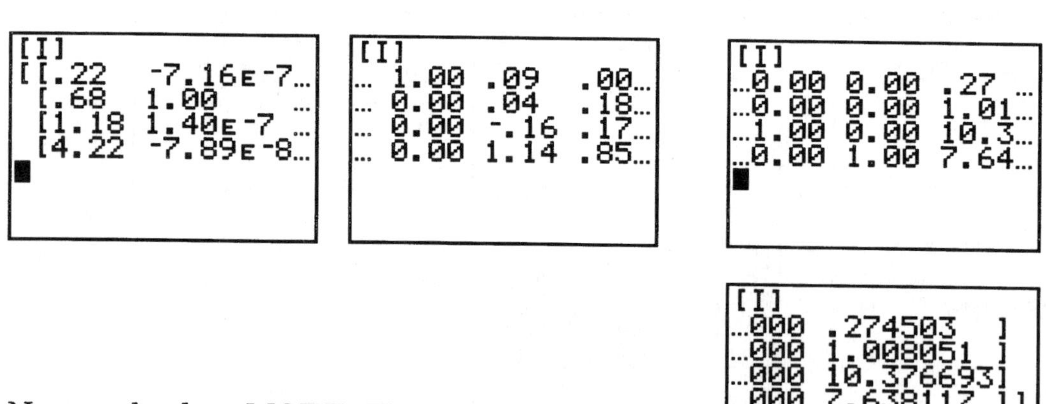

*row+(4.70282,[H],2,4)→[I]

Now, go back to MODE Float 6 and scroll to the final column to read the answers. Use MATRX NAMES to put the final matrix [I] on the Home screen. Record the results and identify the solution as described in the text.

4.4 The Simplex Method: Solving General Linear Programming Problems – TI-83 Plus

This section has no new graphing calculator techniques.

4.5 The Simplex Method and Duality (Optional) – TI-83 Plus

This section has no new graphing calculator techniques.

You're the Expert – Airline Scheduling – TI-83 Plus

This section has no new graphing calculator techniques.

CHAPTER 5

THE MATHEMATICS OF FINANCE

TI-83 PLUS

```
N=24.00
I%=4.75
PV=-2000.00
PMT=0.00
▪FV=2655.06
P/Y=4.00
C/Y=4.00
PMT:END BEGIN
```

5.1 Simple Interest – TI-83 Plus

This section has no new calculator skills. All computations can be done on the Home screen.

5.2 Compound Interest – TI-83 Plus

Compound interest problems can be calculated on the Home screen as well as by using the Finance features of the TI-83 Plus. This section will demonstrate both techniques.

Example 1 - *Savings Accounts*

Let's begin with Example 1 *Savings Accounts*. The problem is stated as follows: *In March, 1994, the Bank of Nova Scotia was paying 4.75% interest on savings accounts. If the interest is compounded quarterly, find the future value of a $2,000 deposit in six years. What is the total interest paid over the period?*

To do this on the calculator, you must first identify all the parts of the problem you know, namely, that the principal PV is $2000, the compound interest rate r is 0.0475, the frequency of compounding is quarterly, so $m=4$, and the number of years t is 6. Next, identify that you are looking for future value FV. Because you are looking for future value of a compound interest

problem you must use the formula $FV = PV\left(1+\dfrac{r}{m}\right)^{mt}$. It is always advisable

to write the original formula as the first step of your solution, then fill in the values of what you know, and finally to work out the result.

The formula $FV = PV\left(1+\dfrac{r}{m}\right)^{mt}$ yields $FV = 2000\left(1+\dfrac{0.0475}{4}\right)^{4\times6}$.

On the Home screen

To do this computation on the Home screen, input

$2000(1+0.0475/4)^{(4*6)}$

Press $\boxed{\text{ENTER}}$ to see the result 2655.062703 or 2655.06.

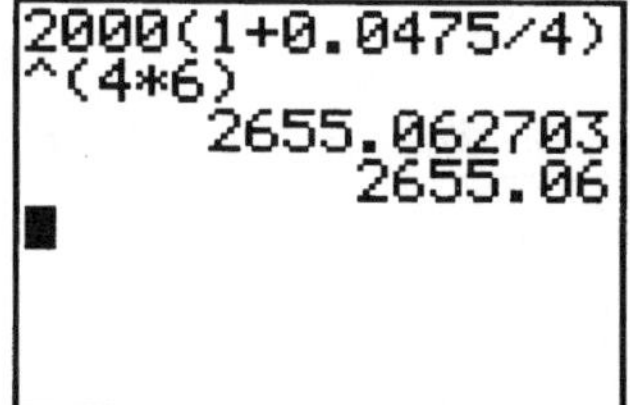

Set MODE to the word Float to get the results as displayed in the text, or use Float 2 since we are working with money. Press CLEAR 2nd MODE (QUIT) to return to the Home screen.

Using Finance features

To do this with the Finance features of the TI-83 Plus, press the blue key, APPS, and select 1:Finance.

Next, select CALC 1:TVM Solver.

Enter the values

N=24 or 4x6	number of times of compounding periods over 6 years;
I%= 4.75	actual percent, not decimal form of the percent;
PV= -2000	Present Value
PMT=0	There are no payments, set PMT to 0;
FV=0	Future Value to find can be any value, say 0;
P/Y=4	4 payments per year;
C/Y=4	compounded 4 times per year,
PMT: END or BEGIN	not used in this example.

Once all of the above is input onto this screen, move the cursor to FV=0, then press ALPHA ENTER (SOLVE). This returns the value FV=2655.06. Record the value.

```
N=24.00
I%=4.75
PV=-2000.00
PMT=0.00
■FV=2655.06
P/Y=4.00
C/Y=4.00
PMT:END BEGIN
```

Press CLEAR 2nd MODE (QUIT) to return to the Home screen.

Evaluate Interest

To evaluate interest you must subtract the original principal from the future value.

```
2655.06-2000
           655.06
```

To do this, go back to the Home screen by pressing 2nd MODE (QUIT).

Input the expression 2655.06-2000 and press ENTER to see the result 655.06.

Example 2 *Zero-Coupon Bonds*

The next example, Example 2 *Zero-Coupon Bonds* computes present value.
The problem is stated as follows: *The Megabucks Corporation is issuing 10-
year zero-coupon bonds. How much would you pay for bonds with a maturity
value of $10,000 if you wish to get a return of 6.5% compounded annually?*

First identify all values:

$$FV=10000, \quad PV=unknown, \quad r=6.5\% \text{ or } 0.065, \quad t=10, \text{ and } m=1.$$

Use the $PV = \dfrac{FV}{\left(1+\dfrac{r}{m}\right)^{mt}}$ formula, and fill in the values before using the

calculator to get the equation $PV = \dfrac{10,000}{\left(1+\dfrac{0.065}{1}\right)^{10\times1}}$.

On the Home screen

To solve this on the Home screen you must be careful to enclose the entire
denominator in parentheses. Input the expression

$$10000/(1+0.065)^\wedge 10*1$$

Press ENTER to see the result 5327.26. Round the
result to hundredths if necessary.

```
10000/(1+0.065)^
10*1
            5327.26
```

Using Finance features

To do this with the Finance features of the TI-83 Plus, press the blue key,
APPS, and select 1:Finance. Select CALC 1:TVM Solver.

Enter the values

    N=10   number of times of compounding periods over 6 years;

    I%= 6.5   actual percent, not decimal form of the percent;

    PV= 0   Present Value we wish to find

```
PMT=0          There are no payments, set PMT to 0;
FV=10000       Future Value
P/Y=1          1 payment per year;
C/Y=1          compounded 1 time per year,
PMT: END or BEGIN          not used in this example.
```

Once all of the above is input onto this screen, move the cursor to PV=0, then press [ALPHA] [ENTER] (SOLVE). This returns the value

$$PV=-5327.260355 \text{ or } -5327.26.$$

Record the value.

Press [CLEAR] [2nd] [MODE] (QUIT) to return to the Home screen.

Example 3 *How Long to Invest*

The next example, Example 3 *How Long to Invest* computes time t that it takes for an investment to grow to a certain amount. The problem is stated as follows: *You have $5,000 to invest at 6% compounded monthly. How long will it take for your investment to grow to $6,000?*

Iidentify all values:

$$FV=6000, \quad PV=5000, \quad r=6\% \text{ or } 0.06, \quad t=\text{unknown}, \text{ and } m=12.$$

Next, solve the equation $FV = PV\left(1+\dfrac{r}{m}\right)^{mt}$ for t by using logs as described in

the text to obtain the equation $\quad t = \dfrac{\log 1.2}{12\log\left(1+\dfrac{0.06}{12}\right)}$.

On the Home screen

To solve this on the Home screen you must be careful to use parentheses as given here.

Input the expression

$$\boxed{\text{LOG}}\ (1.2)/(12\boxed{\text{LOG}}(1+0.06/12))$$

Press $\boxed{\text{ENTER}}$ to see the result 3.046 periods.

Using Finance features

To do this with the Finance features of the TI-83 Plus, press the blue key, APPS, and select 1:Finance. Select CALC 1:TVM Solver.

Enter the values

N=0	unknown number compounding periods
I%= 6	actual percent, not decimal form of the percent;
PV= -5000	Present Value
PMT=0	There are no payments, set PMT to 0;
FV=6000	Future Value
P/Y=12	12 payments per year;
C/Y=12	compounded 12 times per year,
PMT: END or BEGIN	not used in this example.

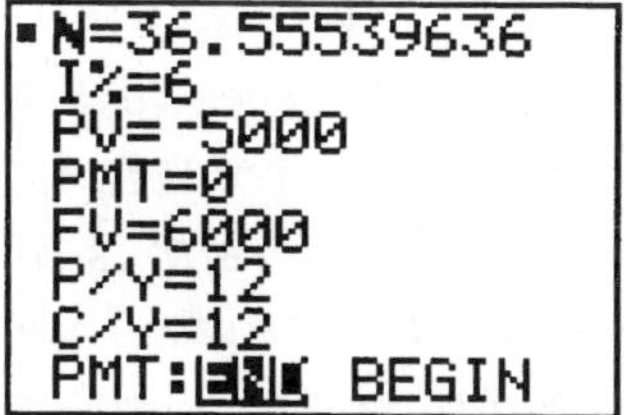

Once all of the above is input onto this screen, move the cursor to N=0, then press $\boxed{\text{ALPHA}}$ $\boxed{\text{ENTER}}$ (SOLVE). This returns the value N=36.55539636 periods per year.   Record the value.

Divide this number N=36.55539636 by 12 to get 3.046 as obtained by using logs on the Home screen.

The result, then, is that it takes a little more than 3 years to grow $5000 to $6000 in this situation.

Press $\boxed{\text{CLEAR}}$ $\boxed{\text{2nd}}$ $\boxed{\text{MODE}}$ (QUIT) to return to the Home screen.

5.3 Annuities, Loans, and Bonds – TI-83 Plus

Annuity, loan, and bond problems can also be calculated on the Home screen as well as by using the Finance features of the TI-83 Plus. Again we will demonstrate both techniques, but I would recommend the Finance features for these examples. In addition this section shows you how to build an amortization schedule.

It is again imperative that you state the formula you wish to use, and to clearly identify all parts of the formula.
The formula for the *Future Value for an Increasing Annuity* is given by

$$ FV = PV(1+i)^n + PMT\left(\frac{(1+i)^n - 1}{i}\right) $$

where
      FV is the future value,
      PV is the present value,
      PMT is the payment at the end of each compounding period,
      r is the annual rate of interest,
      m is the number of times of compounded per year,
      t is the number of years invested,
      i=r/m is the interest paid each period, and
      n=mt is the total number of periods.

Example 1 *Retirement Account*

Example 1 *Retirement Account* computes the future value of an increasing annuity. The problem is stated as follows: *Your retirement account has $5,000 in it and earns 5% interest per year compounded monthly. Every month for the next 10 years you will deposit $100 into the account. How much money will there be in the account at the end of those 10 years?*

To solve this you must first identify all values:
FV=unknown, PV=5000, PMT=100, r=5% or 0.05, m=12, and t=10.
Find i=.05/12 and n=120.

Evaluate the formula $\quad FV = 5000(1+\dfrac{0.05}{12})^{120} + 100\left(\dfrac{(1+\dfrac{0.05}{12})^{120}-1}{\dfrac{0.05}{12}}\right).$

On the Home screen

To solve this on the Home screen you must be careful to use the parentheses as follows. Input the expression

$$5000(1+(0.05/12))\wedge120+100(((1+0.05/12)\wedge120)-1)/(0.05/12)$$

Press [ENTER] to see the result 23763.28. Round the result to hundredths if necessary.

Using Finance Features

To do this with the Finance features of the TI-83 Plus, press the blue key, `APPS`, and select `1:Finance`. Select `CALC 1:TVM Solver`.

Enter the values

`N=120`	number compounding periods
`I%= 5`	actual percent, not decimal form of the percent;
`PV= -5000`	Present Value
`PMT=-100`	Amount of payment
`FV=0`	unknown Future Value
`P/Y=12`	12 payments per year;
`C/Y=12`	compounded 12 times per year,
`PMT: END or BEGIN`	not used in this example.

Once all of the above is input onto this screen, move the cursor to `FV=0`, then press [ALPHA] [ENTER] (`SOLVE`). This returns the value `FV =23763.27543`. Record the value.

Round the result to `FV =23763.28`.

Example 2 *Education Fund*

Example 1 *Education Fund* computes the payment of an increasing annuity.
The problem is stated as follows:
*Tony and Maria have just had a son, Jose Phillipe. They establish an account
to accumulate money for his college education, in which they would like to
have $100,000 after 17 years. If the account pays 4% interest per year
compounded quartery, and they make deposits every quarter, how large must
each deposit be in order for them to reach their goal?*

To solve this you must first identify all values:
`FV=100000, PV=0, PMT=unknown, r=4%, m=4,` and `t=17`.

On the Home screen

To find the amount of payment on the Home screen

input $100000\left(\dfrac{0.01}{1.01^{68}-1}\right)$ as

`100000(0.01)/(1.01^68-1)`

```
100000(0.01)/(1.
01^68-1)
             1033.89
```

Press ENTER to see the result `1033.89`. Round the
result to hundredths if necessary.

Using Finance Features

To do this with the Finance features of the TI-83 Plus, press the blue key,
`APPS`, and select `1:Finance`. Select `CALC 1:TVM Solver`.

Enter the values

`N=68`	number compounding periods
`I%= 4`	actual percent, not decimal form of the percent;
`PV= 0`	Present Value
`PMT=0`	unknown amount of payment
`FV=100000`	Future Value
`P/Y=4`	4 payments per year;
`C/Y=4`	compounded 4times per year,
`PMT: END`	End in this example.

Once all of the above is input onto this screen, move the cursor to `PMT=0`,
then press ALPHA ENTER (`SOLVE`). This returns the value `PMT=-1033.8885`.
Record the value.

Round the result to `PMT=1033.89`.

Examples 3 and 4

Practice the Finance features technique on Examples 3 and 4 in the text.

Example 7 *Amortization Schedules*

An amortization schedule is a table that tells the details about paying a loan. An amortization table contains the number of the month of payment, the amount of the Principal still to be paid (Outstanding Principal) at the end of each payment, the dollar amount of each payment towards the Principal, and the interest paid on a each payment.

The amortization schedule presented in Example 7 is based on Example 6 *Home Mortgages* which states: *Marc and Mira are buying a house, and have taken out a 30 year, $90,000 mortgage at 8% per year. What will their monthly payments be?* Example 7 asks: *How much interest will Marc and Mira pay in the first year of their mortgage?*

The Finance features and table feature of the TI-83 Plus combined allow you to create an amortization table. Here is how to do it.

In Example 6 you found that Marc and Mira's monthly payment is $660.39.

Input these values into the `TVM Solver` window.

`N=360, I%=8, PV=$90,000, PMT=0, FV=0,`
`P/V=12, and C/V=12, END`. Move the cursor to
`PMT=0` and press [ALPHA] [ENTER] (`SOLVE`).

These values will be used to compute the table.

Press [2nd] [MODE](`QUIT`) to return to the Home screen.

Next, input the three functions $Y_1=\texttt{bal(X,2)}$, $Y_2=$ `-∑Prn(X,X,2)`, and $Y_3=$ `-∑Int(X,X,2)`. These three functions are found in the `Finance` menu.

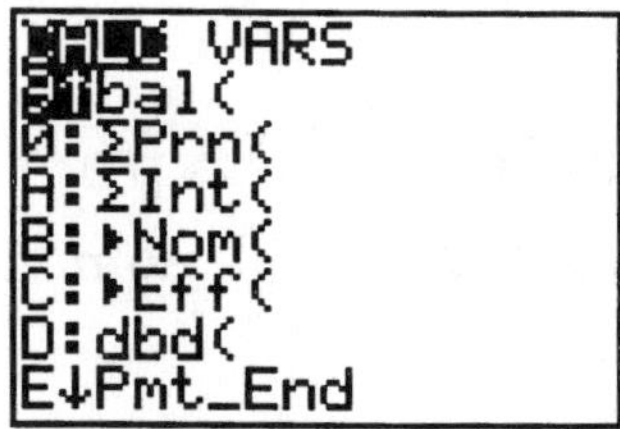

Press Y= to input the functions into the Y= menu.

To input $Y_1 =$ bal(X,2), as Y_1, position the cursor to Y_1, then press APPS. Select 1:Finance, CALC 9:bal(. This puts bal(on Y_1. Finish the function. This rounds the outstanding principal to two places.

To input $Y_2 = -\Sigma$Prn(X,X,2), as Y_2, position the cursor to Y_2, then press APPS. Select 1:Finance CALC 0:ΣPrn(. This puts ΣPrn(on Y_2. Finish the function . This rounds the payment on principal per month to two places.

To input $Y_3 = -\Sigma$Int(x,x,2) as Y_3, position the cursor to Y_3, then press APPS. Select 1:Finance CALC A: ΣInt(. This puts ΣInt(on Y_3. Finish the function . This rounds the amount of interest paid per month to two places.

```
Plot1  Plot2  Plot3
\Y1█bal(X,2)
\Y2█-ΣPrn(X,X,2)

\Y3█-ΣInt(X,X,2)

\Y4=█
\Y5=
```

Next, set the table using 2nd WINDOW (TBLSET) as shown.

```
TABLE SETUP
 TblStart=0
 ΔTbl=1
Indpnt:  Auto  Ask
Depend:  Auto  Ask
```

Finally, press 2nd GRAPH (TABLE) to view the table. Notice that you can only view two functions on one screen, so you must scroll to the right to see the third function.

Column 1 gives the month,
Column 2 gives the outstanding principal,
Column 3 gives the payment on principal, and
Column 4 gives the interest payment.

```
 X    │ Y1     │ Y2
──────┼────────┼───────
 0    │ 90000  │ ERROR
 1    │ 89940  │ 60.39
 2    │ 89879  │ 60.79
 3    │ 89818  │ 61.2
 4    │ 89756  │ 61.61
 5    │ 89694  │ 62.02
 6    │ 89632  │ 62.43
──────┴────────┴───────
X=6
```

```
 X    │ Y2     │ Y3
──────┼────────┼────────
 0    │ ERROR  │ ERROR
 1    │ 60.39  │ 600
 2    │ 60.79  │ 599.6
 3    │ 61.2   │ 599.19
 4    │ 61.61  │ 598.78
 5    │ 62.02  │ 598.37
 6    │ 62.43  │ 597.96
──────┴────────┴────────
Y3=597.96
```

You're the Expert – Saving for College

This section has no new calculator skills.

CHAPTER 6

SETS AND COUNTING

TI-83 PLUS

6.1 Set Operations – TI-83 Plus

This section has no new calculator skills.

6.2 Cardinality – TI-83 Plus

This section has no new calculator skills.

6.3 The Addition and Multiplication Principles – TI-83 Plus

This section has no new calculator skills.

6.4 Permutations and Combinations – TI-83 Plus

This section asks you to evaluate permutations and combinations. The
graphing calculator has both the factorial notation, !, and the built in
formulas for permutations and combinations. In this section, you will see
how to evaluate the formula yourself on the Home screen using !, and also
how to use the built in formulas nPr and nCr. Let's use the quick examples
to demonstrate these operations.

Permutations on the Home Screen using factorial notation

Recall that the *number of permutations of n items taken r at a time* is given
by the formula

$$P(n,r) = \frac{n!}{(n-r)!}$$

The Quick Example of permutations asks you to
evaluate the number of permutations of 6 items
taken 2 at a time, P(6,2). The factorial notation
is found in the MATH PRB menu.

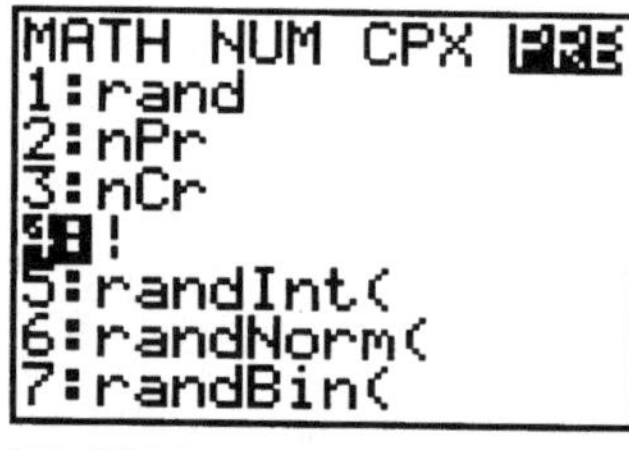

First note that n=6 and r=2.

The permutation formula becomes $P(6,2) = \dfrac{6!}{(6-2)!}$

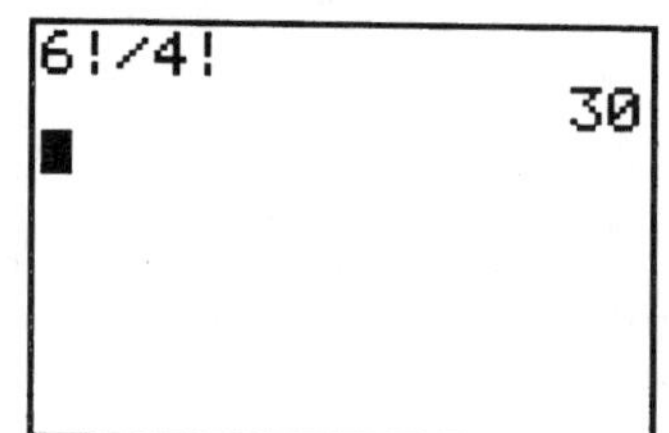

or $P(6,2) = \dfrac{6!}{4!}$

Input 6!/4!. To input the ! sign, press MATH and cursor over to PRB.
Select 4:! to put the ! on the Home screen. Press ENTER to see the result 30.

Permutations on the Home Screen using nPr

The Quick Example of permutations asks you to evaluate the number of
permutations of 6 items taken 2 at a time, P(6,2). The permutation formula
$P(n,r) = \dfrac{n!}{(n-r)!}$ is found in the MATH PRB menu as
nPr.

Again note that n=6 and r=2, so we wish to input
6 nPr 2 on the Home screen.

To input 6 nPr 2, first input the 6.

Press [MATH] and cursor over to PRB. Select 2:nPr to
put the nPr on the Home screen.

Input the 2.

Press [ENTER] to see the result 30.

Combinations on the Home Screen using factorial notation

Recall that the *number of combinations of n items taken r at a time* is given
by the formula

$$C(n,r) = \frac{P(n,r)}{r!} \quad \text{or} \quad C(n,r) = \frac{n!}{r!(n-r)!}$$

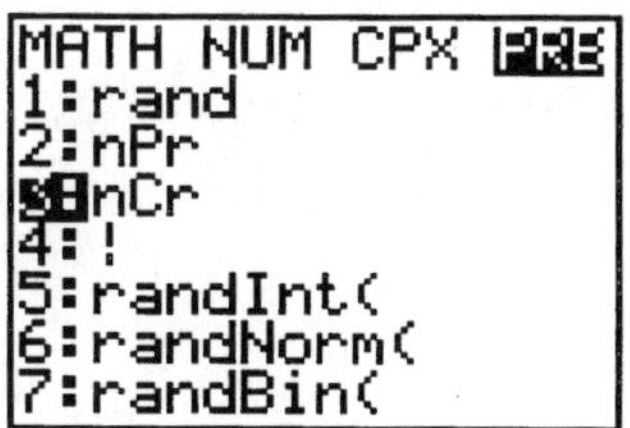

The Quick Example of combinations asks you to
evaluate the number of combinations of 6 items taken
2 at a time, C(6,2). The factorial notation is found
in the MATH PRB menu.

First note that n=6 and r=2.

The combination formula becomes $C(6,2) = \dfrac{6!}{2!(6-2)!}$ or $C(6,2) = \dfrac{6!}{2!4!}$

Input $6!/(2!4!)$. To input the factorial sign, !, press [MATH] and cursor over to PRB. Caution: You *must* use parentheses for the denominator; if you don't, you get an answer of 8640, not 15.

Press [ENTER] to see the result 15.

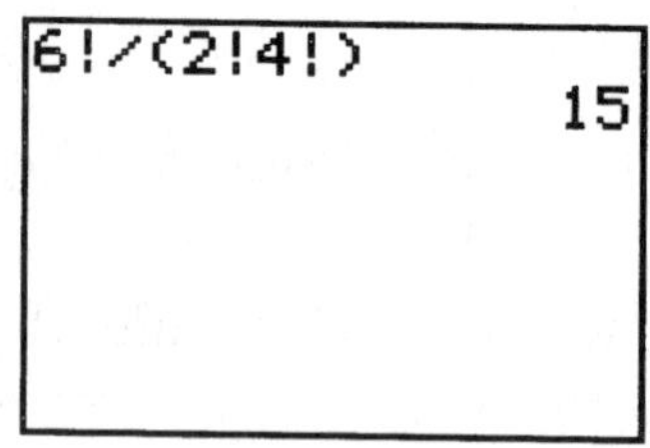

Combinations on the Home Screen using nCr

The Quick Example of combinations asks you to evaluate the number of combinations of 6 items taken 2 at a time, C(6,2). The combination formula $C(n,r) = \dfrac{n!}{r!(n-r)!}$ is found in the MATH PRB menu as nCr.

Again note that n=6 and r=2, so we wish to input 6 nCr 2 on the Home screen.

To input 6 nCr 2, first input the 6.

To input nCr, press [MATH] and cursor over to PRB. Select 3:nCr.

Input the 2.

Press [ENTER] to see the result 15.

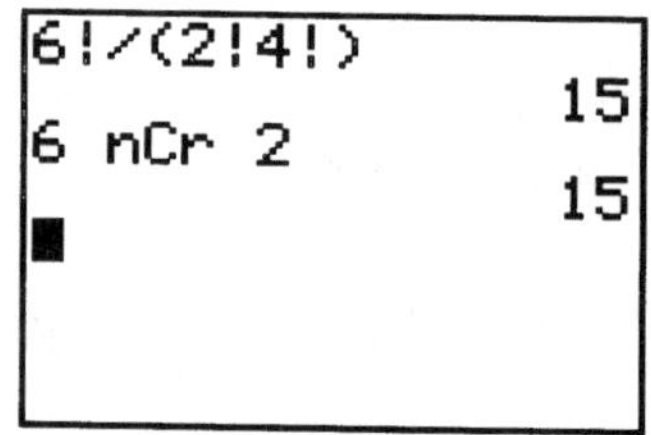

You're the Expert – Designing a Puzzle
This section has no new calculator skills.

CHAPTER 7

PROBABILITY

TI-83 PLUS

```
MATH NUM CPX PRB
1:rand
2:nPr
3:nCr
4:!
5▌randInt(
6:randNorm(
7:randBin(
```

7.1 Sample Spaces and Events – TI-83 Plus

This section has no new calculator skills.

7.2 Experimental Probability – TI-83 Plus

In this section you will see how to generate random numbers to create a sample space. There are two random number generators found on the TI-83 Plus that are useful to you in this section, namely `rand` and `randInt(`. Both can be found in the `Math  PRB` menu. Also, sometimes we wish to round a number to a certain number of decimal places using `round`, found in the `Math  Num` menu.

Let's start by looking at the syntax for these commands.

> `round (x, 0)` where x is the value or function to be rounded, and the number 0 rounds the number to 0 decimal places. `Round` is found by pressing [MATH] `NUM 2:round`.
>
> `rand` generates a number between 0 and 1. `rand` is found by pressing [MATH] `PRB 1:rand`.
>
> `randInt (1,100)` generates a random integer between two given values, say between 1 and 100. `randInt` is found by pressing [MATH] `PRB 5:randInt`.

There are three kinds of random number problems discussed in the text:

1) Coin problems whose results are heads or tails.
   For coin problems we can denote success, such as getting a head, with a 0 or 1. Since random number generators usually produce a number between 0 and 1, we can use two functions, namely round and rand together as `round(rand,0)` to produce a list of 0's and 1's.

2) Dice problems whose results are the numbers 1, 2, 3, 4, 5 or 6 use the command `randInt (1,6)`.

3) Percentage problems whose numbers are between 1 and 100 use the command `randInt (1,100)`. Record the results to count how many entries fall within a given range, say, between 1 and 20.

Operations on Lists

Operations such as sum, cumulative sum, sorting, etc. can be performed on the Home screen using the List menu. To get into the list menu press [2nd] [STAT] (LIST).

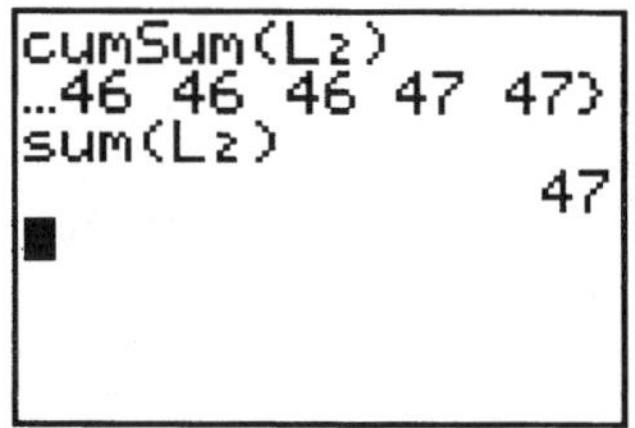

To see the possible operations on lists, cursor to the word OPS. The operation 6:cumSum continuously adds the previous entry to the current entry.

When you cursor over to MATH, you find 5:sum(. This operation finds the sum between certain elements of a list, such as the sum of the first ten entries, or the sum of the first fifty, or the sum of the entries between 11 and 20.

Creating lists

There are two methods for generating the lists on the TI-83 Plus. One is by using the TABLE feature of a function to be graphed. The other method uses lists created using STAT EDIT. These two methods are described in the solution of the coin problem presented below.

Coin problems

Let's use Example 3 - *Simulating Experiments with Random Numbers* as stated in the text:

Use a simulated experiment to check the following.
(a) The theoretical probability of heads coming up in a toss of a fair coin approaches 1/2 as the number of trials gets large.
(b) The probability of heads coming up in two consecutive tosses of a fair coin approaches 1/4 as the number of trials gets large.

The experiment for part (a)

The experiment as described in the text for part (a) consists of generating 100 random numbers between 0 and 1, and then rounding them to the nearest whole number, either Ø or 1. A result of Ø represents *not* getting a head, and a result of 1represents getting a head.

Next, count the number of heads, i.e. the number of 1's that occur after 1Ø, 2Ø, and 1ØØ trials.

Finally, calculate the probabilities for each case.

There are two methods that can be used on the graphing calculator: one method is to create a table in the Y= menu, then count the number of 1's by hand, and calculate the probabilities. The second method uses two lists: the numbers from 1 to 1ØØ and the randomly generated numbers. Sums can be calculated on the second list to count the number of times 1 occurs over 1Ø, 2Ø, and 1ØØ trials.

Method 1 - Using a Table For Example 3a.

First enter the function Y1= round(rand,Ø).

Press the [Y=] key. Position the cursor on Y1.

Press [MATH] NUM 2:round

Then press [MATH] PRB 1:rand [,] [0] [)]

```
Plot1  Plot2  Plot3
\Y1■round(rand,0
)
\Y2=
\Y3=
\Y4=
\Y5=
\Y6=
```

Next, create a table generated by Y_1. Press [2nd] [WINDOW] (TBLSET) to begin table at 1 and step by 1 as shown on the screen.

```
TABLE SETUP
 TblStart=1
 ∆Tbl=1
Indpnt:  Auto  Ask
Depend:  Auto
```

Press [2nd] [GRAPH] (TABLE) to display two columns and record the columns on your paper. Scroll through the table to see x between 1 and 100.

X	Y1	
1	1	
2	1	
3	0	
4	1	
5	1	
6	0	
7	1	

X=1

X	Y1	
8	1	
9	0	
10	1	
11	0	
12	0	
13	1	
14	1	

X=14

X	Y1	
15	0	
16	0	
17	1	
18	1	
19	1	
20	0	
21	1	

X=21

84

Count the number of 1's in the first 10, first 20, and then all 100 trials. Record the results. Note that answers will vary because of the random number generator.

Evaluate the probabilities.

$$P(H) \text{ in the first 10 trials} = \frac{\#1's}{10}$$

$$P(H) \text{ in the first 20 trials} = \frac{\#1's}{20}$$

$$P(H) \text{ in the first 100 trials} = \frac{\#1's}{100}$$

Method 2 - Using Lists For Example 3a

You can use STAT EDIT to create two lists, L_1 and L_2. Let L_1 be the numbers from 1 to 100. Let L_2 be the values of Y_1= round(rand,0), the randomly generated values of 0's and 1's.

The advantage of using lists is that the calculator will count the number of 1's for you using the list operation sum.

To create L_2, first enter the function Y_1= round(rand,0) as described above.

Enter the lists into STAT EDIT. Press the [STAT] key, the cursor should be on the word EDIT. Select 1:Edit.

In the L_1 column input the numbers 1 to 100. You can do this by hand, or you can use the command seq(X,X,1,100)→L_1.

To use the sequence command, press [2nd] [STAT] (LIST).

Move the cursor to OPS and select 5:seq(.

Enter the rest of the command
seq(X,X,1,100)→L₁.

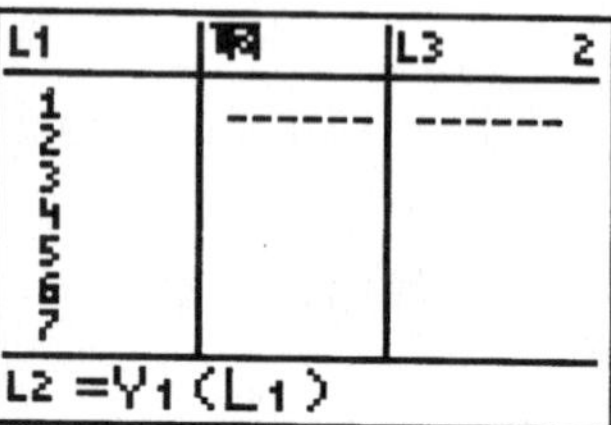

Use the $\boxed{\text{STO▶}}$ key to store the list into L_1.

In the L_2 column, move the cursor to the name L_2.

Enter $Y_1(L_1)$. To do this, press the $\boxed{\text{VARS}}$ key.
Move the cursor to Y-VARS, Select 1:Function.
Move the cursor to 1: Y_1 and press $\boxed{\text{ENTER}}$. Use
$\boxed{\text{2nd}}$ $\boxed{1}$ (L_1) to enter L_1.

Press $\boxed{\text{ENTER}}$ to see the new list. Scroll through the list with the down arrow to
see the entire list. Note that your list may be
different from the one shown because we are
generating random numbers.

Record the results on your paper.

Press $\boxed{\text{2nd}}$ $\boxed{\text{MODE}}$ (QUIT) to return to the Home
screen.

Next, count the number of 1's in the first 10, first 20, and then all 100 trials.
Record the results. Note that answers will vary because of the random
number generator. Use the list operation sum found in the List MATH
menu. The syntax for sum is
        sum (list name, start entry, stop
entry).

To find the sum of the number of heads in the first
ten trials press $\boxed{\text{2nd}}$ $\boxed{\text{STAT}}$ (LIST). Cursor to the
word MATH, select 5:sum(.

Enter the rest of the command to see
sum(L₂,1,10) on the Home screen. Press $\boxed{\text{ENTER}}$ to
see the result. Note that your result and someone

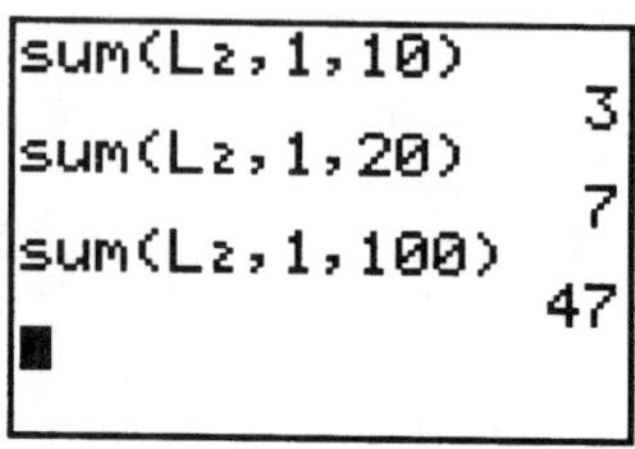

else's result will be different because different random numbers were generated.

Repeat the line using [2nd] [ENTER] (ENTRY). Change the command to read sum(L₂,1,20). Press [ENTER] to see the result.

Repeat and change the line to see read sum(L₂,1,100). Press [ENTER] to see the result.

Evaluate the probability for each case by hand or with the calculator. Here are the results from the screens shown above.

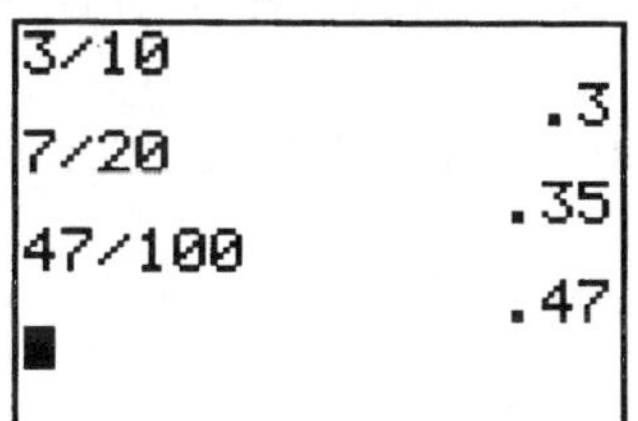

    P(H) in the first 10 throws is 3/10 or 0.3.
    P(H) in the first 20 throws is 7/20 or 0.35.
    P(H) in the first 100 throws is 47/100 or 0.47.

Now let's do Example 3b- *Simulating Experiments with Random Numbers* as stated in the text:

    Use a simulated experiment to check the following.
    The probability of heads coming up in two consecutive tosses of a fair coin
    approaches 1/4 as the number of trials gets large.

The Experiment

The experiment in 3b is to create a list of 100 random numbers of 0's and 1's where 1 is the success of getting two consecutive heads, and 0 is *not* getting two consecutive heads.

The same two methods can be used on the graphing calculator to create the lists: to create a table in the Y= menu, then count the number of 1's by hand, and to create two lists: the numbers from 1 to 100 and the randomly generated numbers and use List sum to count the number of 1's.

Calculate the probabilities of success, the number of 1's in 10, 20, 50, and 100 trials.

Method 1 - Using a Table For Example 3b.

Let's use the greatest integer function as explained in the text.

First enter the function as shown.

$Y_2=\text{int}(0.5*(\text{round}(\text{rand},0)+\text{round}(\text{rand},0)))$.

Press the [Y=] key. Position the cursor on Y_2.

Press [MATH] NUM 5:int(.

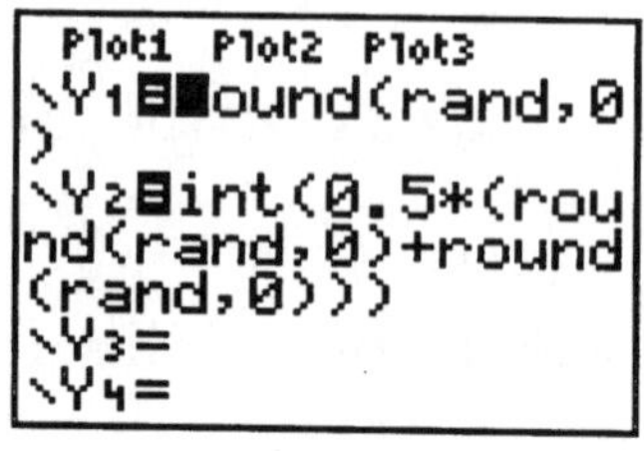

Then enter the .5*(.

Press [MATH] NUM 2:round(.

Press [MATH] PRB 1:rand [,] [0] [)].

Continue to enter the rest of the equation as shown.

Next, look at the table generated by Y_2.

Be sure that the settings for the table are as
shown on the screen.

Press [2nd] [GRAPH] (TABLE) to display two
columns and record the columns on your paper.
Scroll through the table to see $Y_2(x)$ between 1
and 100. Record the results.

Count the number of 1's in the first 10, first 20,
and then all 100 trials. Record the results.
Note that answers will vary because of the random number generator.

Evaluate the probabilities.

$$P(H) \text{ in the first 10 trials} = \frac{\#1's}{10}$$

$$P(H) \text{ in the first 20 trials} = \frac{\#1's}{20}$$

$$P(H) \text{ in the first 50 trials} = \frac{\#1's}{50}$$

$$P(H) \text{ in the first 100 trials} = \frac{\#1's}{100}$$

Method 2 – Using Lists For Example 3b

You can use `STAT EDIT` to create two lists, L_1 and L_3. L_1 is already entered. Here are the steps to enter the lists of L_3.

To create L_3, first enter the function

```
Y₂=int(0.5*(round(rand,0)+round(rand,0)).
```

```
int(  is found in [MATH] NUM 5:int(
round(  is found in [MATH] NUM 2:round(
rand  is found in [MATH] PRB 1:rand.
```

To enter the lists into `STAT EDIT`, press the [STAT] key. The cursor should be on the word `EDIT`.

In the L_1 column input the numbers 1 to 100, if not already done.

In the L_3 column, move the cursor to the name L_3.

Enter L_3 as $Y_2(L_1)$. To do this, press the [VARS] key. Move the cursor to `Y-VARS, 1:Function` and press [ENTER]. Move the cursor to `2: Y₂` and press [ENTER]. Use [2nd][1] (L_1) to enter L_1.

Press [ENTER] to see the new list. Scroll through the list with the down arrow to see the entire list.

Record the results on your paper.

Next, count the number of 1's in the first 10, first 20, 50, and then all 100 trials. Record the results. Note that answers will vary because of the random number generator. Use the list operation `sum` found in the `List MATH` menu. The syntax for sum is

```
sum (list name, start entry, stop entry).
```

To find the sum of the number of heads in the first
ten trials press [2nd] [STAT] (LIST). Cursor to the
word MATH, select 5:sum(.

Enter the rest of the command to see
sum(L₂,1,10) on the Home screen. Press [ENTER] to see the result. Note that
your result and someone else's result will be different because different
random numbers were generated.

Repeat the line using [2nd] [ENTER] (ENTRY).
Change the command to read sum(L₂,1,20).
Press [ENTER] to see the result.

Repeat the line using [2nd] [ENTER] (ENTRY). Change the command to read
sum(L₂,1,50). Press [ENTER] to see the result.

Repeat and change the line to see read
sum(L₂,1,100). Press [ENTER] to see the result.

Evaluate the probability for each case by hand or
with the calculator. Here are the results from the
screens shown above.
   P(H) in the first 10 throws is 2/10 or 0.2.
   P(H) in the first 20 throws is 5/20 or 0.25.
   P(H) in the first 50 throws is 13/50 or 0.26.
   P(H) in the first 100 throws is 23/100 or 0.23.

Dice problems

For six sided die problems use the random integer
command in Y₃=randInt(1,6). Input Y₃ into the
Y= menu as shown.

randInt(is found in [MATH] PRB 5:randInt(.

The Table Method

Use the table method to generate a list of results as shown. Record the results.

Count the number of 1's, 2's, 3's, etc. as needed in the problem by hand. For example, there are 2 fives on this screen.

Evaluate the probabilities as needed. For example, the probability of getting a five from this screen is 2/7 or about 28.6%.

The List Method

To use the list method, enter L_4 as $Y_3(L_1)$. Press ENTER to generate the list.

Count the number of 1's, 2's, 3's, etc. as needed in the problem. For example, there are 3 ones on this screen.

Evaluate the probabilities as needed. For example, the probability of getting a one on this screen is 3/7.

Percentage problems such as 7.2 Exercise 37

Percentage problems, whose numbers are between 1 and 100, use the command randInt (1,100). Record the results to count how many entries fall within a given range, say, between 1 and 20.

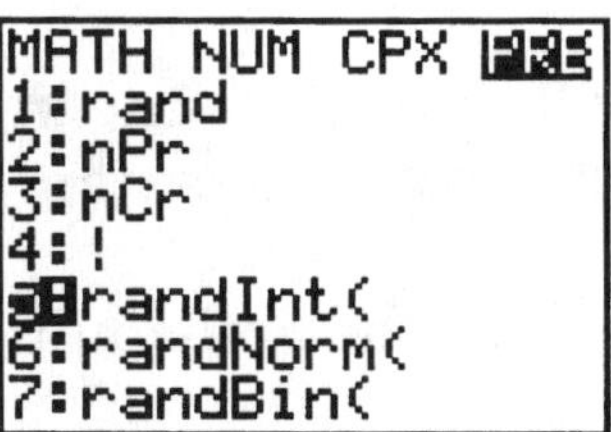

For percentage problems use the random integer command in
Y_4=randInt(1,100).

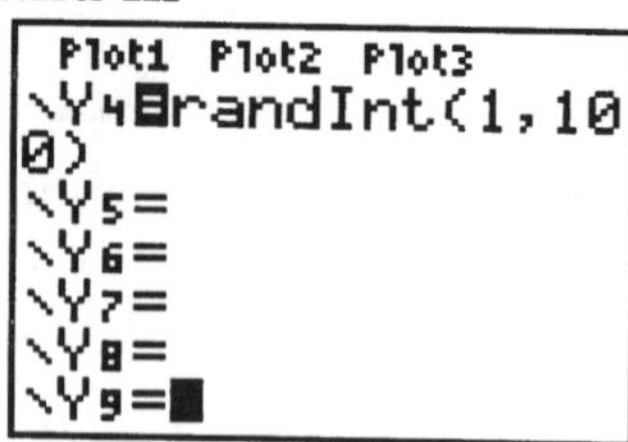

Input Y_4 into the Y= menu as shown.
randInt(is found in [MATH] PRB 5:randInt(.

The Table Method

Use the table method to generate a list of results as shown.

Scroll through x=1 to x=100. Record the results, and
count the number of entries within a certain range,
say between 1 and 20.

 Evaluate the probabilities needed.

The List Method

To use the list method, enter L_5 as $Y_4(L_1)$. Press [ENTER] to generate the list.

Input Y_4=randInt(1,100) into the Y= menu as shown.

Enter $Y_4(L_1)$ as (L_5) as shown. Press [ENTER] to view the list.

Scroll through x=1 to x=100. Record the results, and count the number of
entries within a certain range, say between 1 and 20.

Evaluate the probabilities as needed.

7.3   Empirical Probability – TI-83 Plus

This section has no new calculator skills.

7.4   Probability and Counting Techniques – TI-83 Plus

This section has no new calculator skills. However, you may wish to review combinations that were discussed in Chapter 6.

7.5   Abstract Probability – TI-83 Plus

This section has no new calculator skills.

7.6   Conditional Probability and Independence – TI-83 Plus

This section has no new calculator skills.

7.7   Bayes' Theorem and Applications – TI-83 Plus

This section has no new calculator skills.

7.8   You're the Expert - The Monty Hall Problem – TI-83 Plus

This section has no new calculator skills.

CHAPTER 8

RANDOM VARIABLES
AND
STATISTICS

TI-83 PLUS

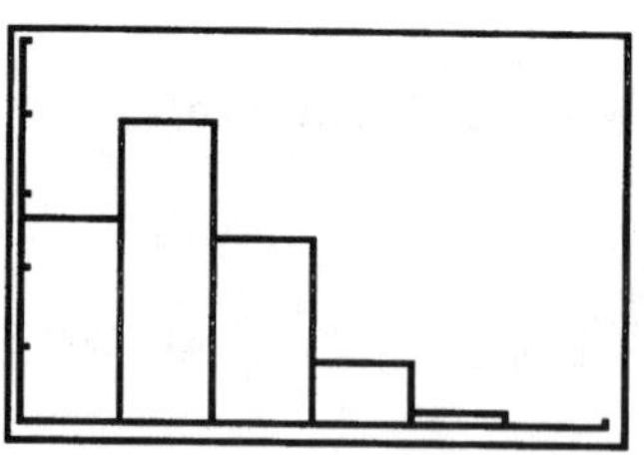

8.1 Random Variables and Distribution - TI-83 Plus

In this section you will see how to draw a histogram on the graphing calculator.

Histograms

Let's use Example 3 - *Empirical Probability Distribution* to demonstrate how to construct a histogram. Example 3 states *Let X be the number of heads in three tosses of a coin. Give the probability distribution of X.*

From the discussion of Example 1 we see that the following table in the text gives the number of heads for each possible outcome.

Outcome	HHH	HHT	HTH	HTT	THH	THT	TTH	TTT
Value of X	3	2	2	1	2	1	1	0

We are told in the text of Example 3 that the random variable X has the values of 0, 1, 2, and 3. This means that we can expect the number of heads on any given toss to be 0, 1, 2, or 3 heads.

Examining the outcome chart above tells that there are eight possible outcomes, so the probability of getting 0 heads, for example, is 1 chance in 8 or 1/8. We examine the remaining outcomes and find the probability distribution to be as follows, as is stated in the text.

x	0	1	2	3
P(X = x)	1/8	3/8	3/8	1/8

To graph this probability distribution as histogram on the TI-83-Plus, do the following you must clear L_1 and L_2, or use an open list.

To clear a list, go into STAT EDIT and select 4:ClrList. This puts ClrList on the Home screen.

```
ClrList L1
               Done
ClrList L2
               Done
■
```

Next enter the name of the list you wish to clear.
The list names are on the keyboard, for example L_1 is 2nd 1 (L_1).

You can also get the name of the list from the List menu, [2nd] [STAT] (LIST) NAMES 1:L$_1$. Press [ENTER] to get the name on the Home screen. Press [ENTER] again to clear the list and see the word Done.

Now enter the random variable values of X into L$_1$. Press the [STAT] key, go into 1:Edit, and enter the values of x. Then enter the probabilities into L$_2$. Note that you may enter the probabilities as fractions, but the calculator displays them as decimals.

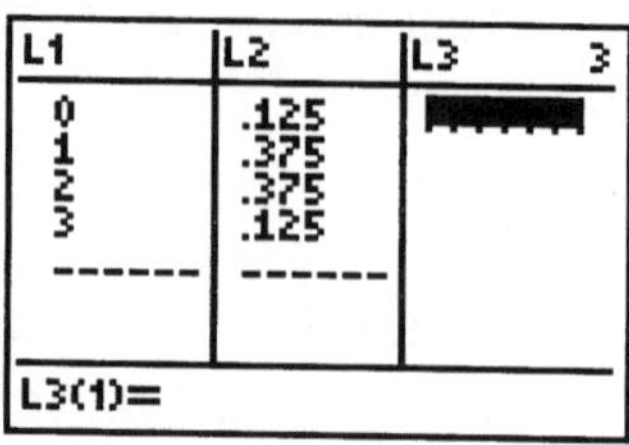

Set the window to Xmin=0, Xmax=4, Xscl=1, Ymin=0, Ymax=0.5, and Yscl=0.1.

Next, set up the Stat Plot, say Plot 1. To do this, press [2nd] [Y=] (STAT PLOT), move the cursor to 1:Plot 1 and press [ENTER].

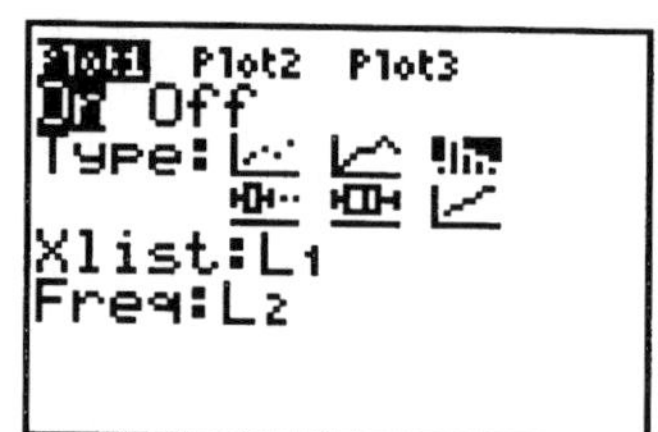

Select On, Type: ▥, xList:L$_1$, Freq: L$_2$. This tells the calculator that you want Plot 1 on, the Type of graph is histogram, XList is coming from L$_1$, and the frequency Freq is coming from L$_2$.

Press [GRAPH] to see the histogram. Explore the graph with [TRACE].

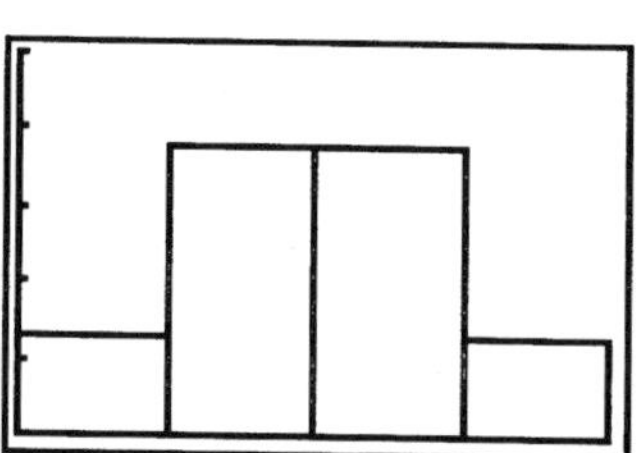

Practice the procedures described above for Example 5 - *Nutrition.*

8.2 (Optional) Bernoulli Trials and Binomial Random Variables - TI-83 Plus

This section discusses how to make a table and a histogram of the probability distribution of a binomial random variable. There are two approaches to making this table. One is to input the formula of this distribution directly into the Y= menu; the other is to use the binomial distribution command found on the calculator, `binompdf(`. To create the histogram, we must put the variable x into L_1 and the probabilities into L_2.

Let's illustrate parts *(a)* and *(b)* of Example 2 *Will You Still Need Me When I'm 64?* The example states: *The probability that a randomly chosen person in the US is 65 or older is approximately 0.2.*
  (a) *What is the probability that, in a randomly selected sample of 6 people, 4 of them will be 65 or older?*
  (b) *If X is the number of people of age 65 or older in a sample of 6, construct the probability distribution of X and plot its histogram.*

Using Y= and the formula

Recall from the text that the formula describing this situation is given by $P(X = x) = C(n,x)p^x q^{n-x}$, where `n=6`, `p=0.2`, and `q=0.8`.

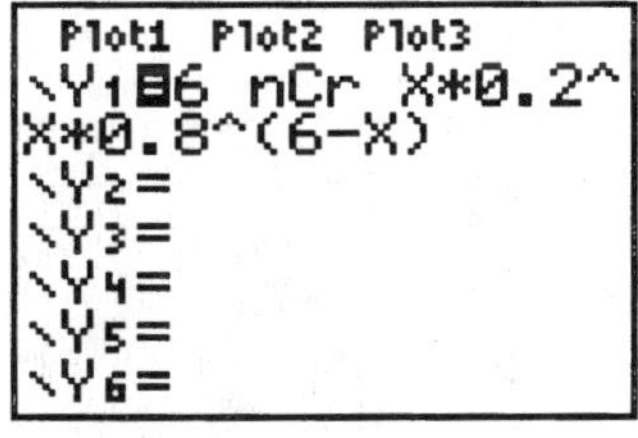

To evaluate the probabilities for the sample space when x = `0,1,2,3,4,5,6` we can input the formula into the Y= menu.

To do this, press the $\boxed{\text{Y=}}$ key, and in Y_1 put the formula $C(n,x)p^x q^{n-x}$ as
`6nCrX*0.2^X*0.8^(6-X)`.

Note that the * is not essential, but it does make the formula more readable. Recall that `C(n,r)` is found using $\boxed{\text{MATH}}$ `PRB 3:nCr`.

Set TABLE SETUP to Tbl Start=0, Δ Tbl=1, Indpnt: Ask, Depend: Auto.

Enter the values X=0 through X=6. Press [ENTER] after each entry.

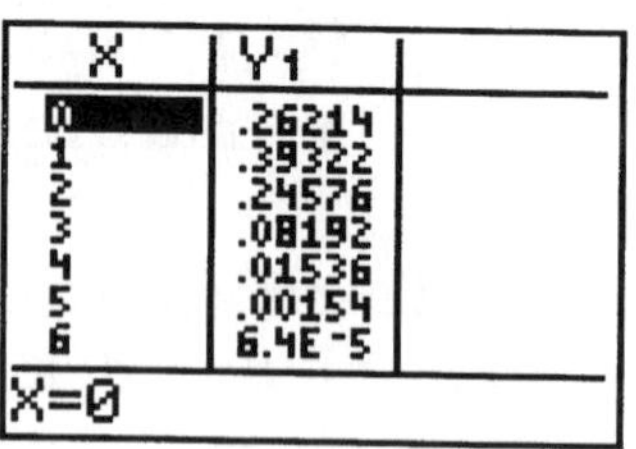

To view the table for this function, press [2nd] [GRAPH] (TABLE).

Using Y= and binompdf(and TABLE

The TI-83 Plus contains a number of common distributions. The binomial distribution is found in the DISTR menu.

To use Y= and binompdf(, go into the Y= menu and position the cursor on Y_2.

Press [2nd] [VARS] (DISTR).

On the DISTR screen, select 0:binompdf(. This puts binompdf(at Y_2.

The syntax for this command is binompdf(n,p,x),

so enter 6,0.2,X to see Y_2= binompdf(6,0.2,X) as shown. Keep both Y_1 and Y_2 turned on to see both tables side by side.

Press [2nd] [GRAPH] (TABLE) to view the tables. Note that the tables are identical.

Plot the Histogram using Y= and Lists

We have constructed the probability distribution in two ways, now we must plot the histogram. Recall from the discussion in 8.1 that a histogram requires the data to come from two of the lists, say L_1 and L_2. Let's enter the X's into L_1 and the probabilities into L_2.

First clear L_1 and L_2 using $\boxed{\text{STAT}}$ EDIT 4:ClrList as described in 8.1.

Next go into $\boxed{\text{STAT}}$ EDIT 1:Edit.

Enter 0 through 6 in L_1.

Move the cursor to the name of L_2. Enter $Y_2(L_1)$. To enter Y_2 , use $\boxed{\text{VARS}}$ Y-VARS 1:Function $\boxed{\text{ENTER}}$ 2: Y_2. Use the keyboard to enter L_1.

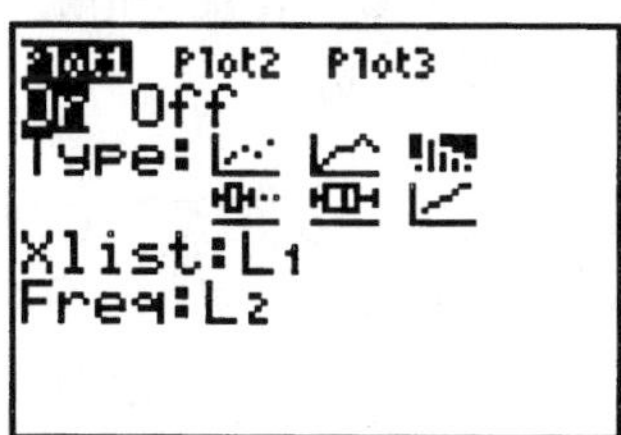

Press $\boxed{\text{ENTER}}$ to view the two lists.

To graph the histogram, press $\boxed{\text{2nd}}$ $\boxed{\text{Y=}}$ (STAT PLOT) and turn on Plot 1. Make the necessary selections, namely to draw a histogram, and take the values of X from L_1 and the values of Y , the Freq, from L_2.

Set the window to Xmin=0, Xmax=6, Xscl=1, Ymin=0, Ymax=0.5, and Yscl= 0.1.

Press $\boxed{\text{GRAPH}}$ to view the histogram.

Use $\boxed{\text{TRACE}}$ to explore the graph.

8.3 Measures of Central Tendency - TI-83 Plus

In this section you will see how to calculate mean, median, and expected
value on the graphing calculator.

The mean

The mean is given by the formula $\bar{x} = \dfrac{\sum_i x_i}{n}$. To evaluate the mean, $\bar{x}$, on
the TI-83 Plus of a set of numbers, say, those in Example 1 *Teen-age
Spending*, you must first enter the values into the list menu, then use the
STAT CALC menu of one variable statistics.

Enter the values {90, 90, 85, 80, 80, 80,
80, 85, 90, 100} into one of the lists, say, L_1.
You may need to clear the list using [STAT] EDIT
4:ClrList.

Next, go into [STAT] CALC 1:1-var Stats. Press
[ENTER] to see 1-Var Stats on the Home screen.

Press [ENTER] again to view the stats. Notice that
$\bar{x}=86$ is the first entry in the list, and that the
number of values in the list is 10, n=10.

Record the mean. Scroll down to see the other information that is
calculated for this set of numbers.

The median

When the numbers in the list are arranged in
ascending order, the median is the "middle" of the
list. The median is the *average* of the two middle
entries of the list when there are an even number
of entries. The median is the *exact middle* of the
list when the number of the entries is odd.

For the set of data of Example 1 *Teen-age Spending*, the median Med=85 is
one of the values in the 1-Var Stats found above.

Expected Value of a Random Variable

The expected value of a random variable as explained in the text is the prediction of the mean of a probability distribution without taking a sample. Expected value is the sum of the products of the x_i's and their corresponding probabilities of a probability distribution. Expected value is given by the formula

$$\mu = E(x) = \sum_i x_i \cdot P(X = x_i).$$

Let's use Example 3 *Sports Injuries* to demonstrate this process. The problem states: *According to historical data, the number of injuries that a member of the Enormous State University women's soccer team will sustain during a typical season is given by the following probability distribution table. If X denotes the number of injuries encountered by a player during the season, compute E(x) and interpret the results.*

Injuries	0	1	2	3	4	5	6
Probability	0.2	0.2	0.22	0.2	0.15	0.01	0.02

To evaluate the expected value of this example on the TI-83 Plus, first enter the values of X into L_1, and the corresponding probabilities into L_2. Then use the sum operation in the list menu to evaluate the expected value.

Clear the lists if necessary. Enter the number of injuries $\{0,1,2,3,4,5,6\}$ into L_1.

Enter the probabilities $\{0.2, 0.2, 0.22, 0.2, 0.15, 0.01, 0.02\}$ into L_2.

Return to the Home screen to evaluate the expected value. Then, go into the List menu, [2nd] [STAT] (LIST) and select MATH 5:Sum(. This puts sum(on the Home screen.

Input the rest of the command to read sum(L_1*L_2). Press [ENTER] to see the result 2.01.

8.4 Measures of Dispersion – TI-83 Plus

This section discusses sample standard deviation s, denoted by Sx on the TI-83 Plus, and population standard deviation σ, denoted by σx on the TI-83 Plus.

Standard deviation

The formula for sample mean is given by $\displaystyle \bar{x} = \frac{\sum_i x_i}{n}$

The formula for sample variance is given by $\displaystyle s^2 = \frac{\left[\sum_{i=1}^{n}(x-\bar{x})^2\right]}{n-1}$.

The formula for standard deviation Sx is given by $s = \sqrt{s^2}$.

Let's use Example 1 *Television Ratings* to illustrate the usage of these measures of dispersion on the TI-83 Plus. The example reads:
The Nielsen ratings of television shows are carefully watched by buyers of television commercials. The following sample represents 8 years of Nielsen Ratings for ABC's Academy Awards Show: { 28, 29, 30, 31, 31, 33, 30, 28 }. Compute the sample mean and standard deviation. What percentage of the scores fall within one standard deviation of the mean? What percentage fall within two standard deviations of the mean?

First enter the ratings into a list, say, L_1. Note that you may need to clear the list first. Then use STAT CALC on one-variable data to find the mean and the standard deviation.

To clear L_1 press [STAT], then select EDIT 4:ClrList.

Press [2nd] [1] to input , L_1 and see ClrList L_1 on the Home screen. Press [ENTER] to see the word Done.

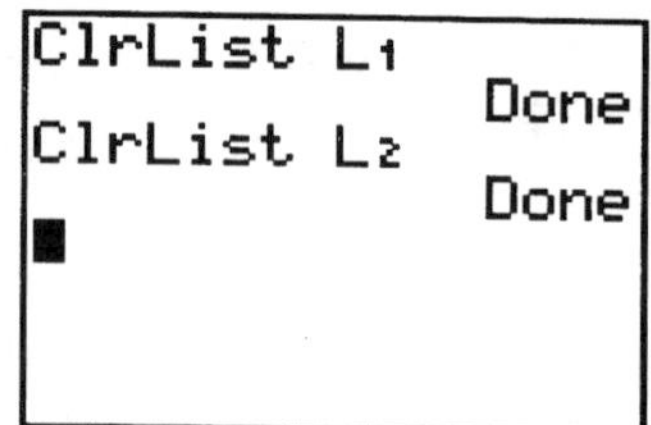

Repeat the clear list process to clear other lists.

Next, input the ratings { 28, 29, 30, 31, 31, 33, 30, 28 } into L_1 by using STAT EDIT 1:Edit.

Note that you can set the MODE to Float 2 to see the stats rounded to two decimal places.

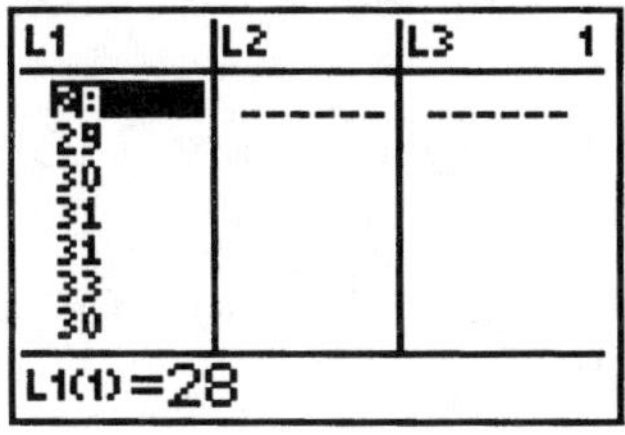

To see the mean and the sample standard deviation Sx, use STAT CALC and select 1:1-Var Stats. This puts 1-Var Stats on the Home screen. Press ENTER to view the stats.

You see that the mean $\bar{x}$ is 30, and the standard deviation Sx is 1.69, rounded to two decimal places. Also note, that n, the number of data points, is 8.

Use the up and down arrows to scroll through the 1-Var Stats calculations.

The remaining parts of this example are calculations that can be done individually by hand or on the Home screen.

Variance

Note that if you wish to compute the sample
variance, $s^2 = \dfrac{\left[\displaystyle\sum_{i=1}^{n}(x-\bar{x})^2\right]}{n-1}$, you can record the
value of the standard deviation found in the STAT CALC 1:1-Var Stats listing, Sx=1.69, and then square it on the Home screen to get 2.86.

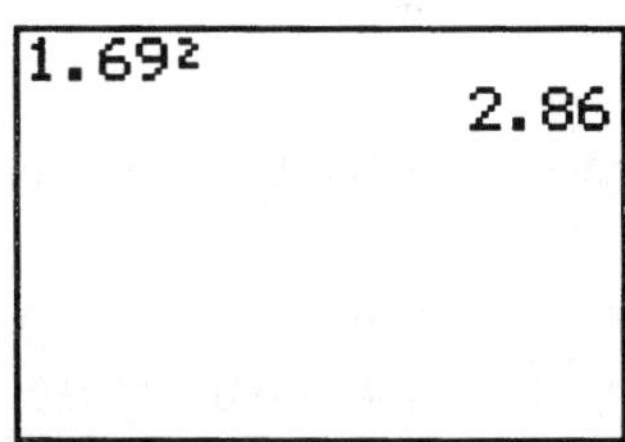

Another way to compute variance is to use the VARS key. To do this, press VARS 5 to select 5:Statistics.

Then choose XY 3: Sx to put Sx on the Home screen.

Press [ENTER] to see 1.69.

Press [x^2] [ENTER] to view the result 2.857142857 or 2.86.

You can also use [x^2] [ENTER] immediately to do this computation in one step as shown.

Variance of a Random Variable

As explained in the text, the population variance is the expected value of $(X-\mu)^2$, which can be written as $E([X-\mu])^2$. So $\sigma^2 = \sum_i (x_i - \mu)^2 P(X = x_i)$,

and the standard deviation is $\sigma = \sqrt{\sigma^2}$. Let's use example 3 to illustrate how to find variance and standard deviation of a random variable on the TI-83 Plus.

Example 3 *Variance of a Random Variable* reads: *Compute the variance and standard deviation for the following probability distribution.*

x	10	20	30	40	50	60
P(X=x)	0.2	0.2	0.3	0.1	0.1	0.1

First evaluate the expected value of this example.

Clear the lists if necessary. Enter the X's {10, 20, 30, 40, 50, 60} into L_1.

Enter the probabilities {0.2, 0.2, 0.3, 0.1, 0.1, 0.1} into L_2.

Press [2nd] [MODE] (QUIT) to return to the Home screen.

To evaluate the expected value, go into the List menu, [2nd] [STAT] (LIST) and select MATH 5:sum(. This puts sum(on the Home screen.

Input the rest of the command to read sum(L₁ * L₂).

On the same line, store the result to M using [STO▸] [ALPHA] [÷] (M).

Press [ENTER] to see the result 30.

Next, evaluate the variance using
$$\sigma^2 = \sum_i (x_i - \mu)^2 P(X = x_i).$$

To evaluate the variance, go into the List menu, [2nd] [STAT] (LIST) and select MATH 5:sum(. This puts sum(on the Home screen.

Input the rest of the command to read sum((L₁-M)^2 * L₂).
Input M using [ALPHA] [÷] (M).

Note that the two left parentheses are essential to get the correct result.

Press [ENTER] to see the result 240.

Note that you could also use the [x²] key instead of the ^2. The command would read sum((L₁-M)^2 * L₂).

Variance and Standard Deviation of a Binomial Random Variable
As explained in the text, the variance of a binomial random variable is given by $\sigma^2 = npq$ and the standard deviation is given by $\sigma = \sqrt{npq}$, where n is the number of independent Bernoulli trials, p is the probability of success and q is the probability of failure.

The expected value μ ,variance σ^2, and standard deviation σ can be evaluated on the Home screen as arithmetic computations. Let's use Example 5 *Interstate Commerce* when n=25, p=0.2 , and q=0.8 .

To compute the expected value $\mu = np$, input 25*0.2 or (25)(0.2) on the Home screen. Press [ENTER] to see the result 5.

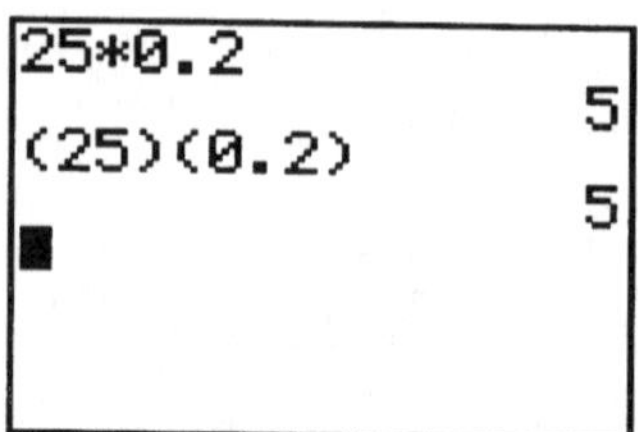

To compute the variance $\sigma = npq$ input (25)(0.2)(0.8) or 25*0.2*0.8 on the Home screen. Press [ENTER] to see the result 4.

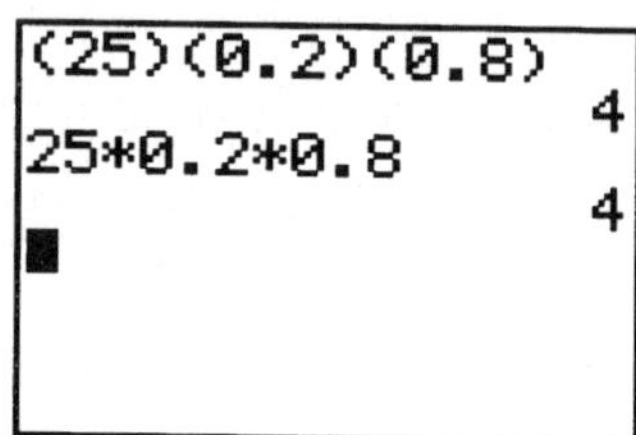

To compute the standard deviation $\sigma = \sqrt{npq}$ on the Home screen from the variance, first input the $\sqrt{\ }$ sign using [2nd] [x^2] ($\sqrt{\ }$) . Next , input Ans using [2nd] [(-)] (ANS) .

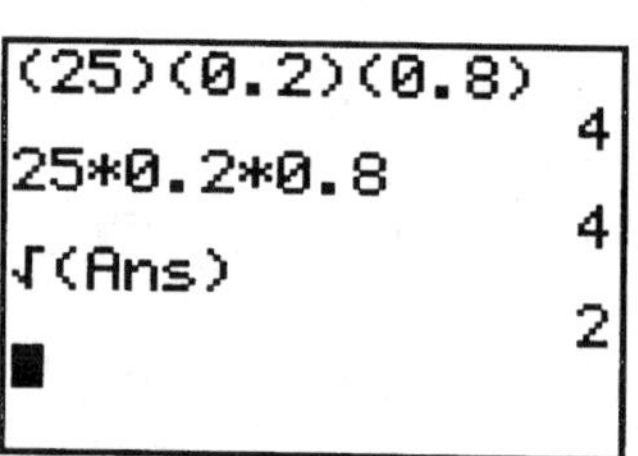

To compute the standard deviation $\sigma = \sqrt{npq}$ on the Home screen directly, first input the $\sqrt{\ }$ sign using [2nd] [x^2] ($\sqrt{\ }$) .

Input the rest of the line to read $\sqrt{\ }$(25*0.2*0.8) or $\sqrt{\ }$((25)(0.2)(0.8)) .

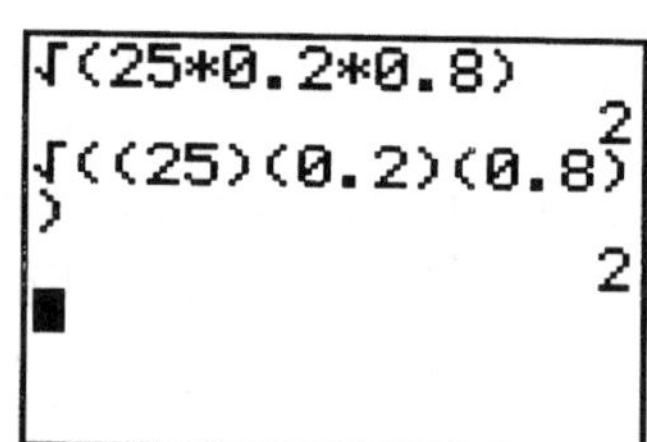

Note that there is a set of parentheses to mark the beginning and end of the radical as well as parentheses to indicate each of the three factors.

Press [ENTER] to see the result 2.

8.5 Normal Distributions – TI-83 Plus

This section discusses the normal distribution. As explained in the text, we find the probability P(a≤Z≤ b) of a standard normal distribution by finding the area under the curve between the vertical lines Z=a and Z=b. On the TI-83 Plus, area under the standard curve is found using in the DIST menu and the command normalcdf(.   The syntax for this command is

```
normalcdf(first Z value, second Z value)
```

Calculating Probabilities Using the Standard Normal Distribution

Let's find the areas under the curve for Example 1 *Standard Normal Distribution.* The example reads:

Let Z be the standard normal variable. Calculate the following probabilities
(a) P(0 ≤ Z ≤ 2.4)
(b) P(0 ≤ Z ≤ 2.43)
(c) P(-1.37 ≤ Z ≤ 2.43)
(d) P(1.37 ≤ Z ≤ 2.43)

Press [2nd] [VARS] (DISTR) to see the distribution menu.

Select DISTR 2:normalcdf(. Press [ENTER] to put this on the Home screen.

Enter the values 0 and 2.4 of Z for part (a) so the command reads normalcdf(0,2.4).

Press [ENTER] to see the result .4918024706 or .4918 rounded to four decimal places. Round the result by hand or set MODE to Float 4. Press [ENTER] again if you change the MODE.

Repeat the line using [2nd] [ENTER] (ENTRY).

Use the arrow keys to edit the command for the next set of values of Z.

Input the commands
```
normalcdf(0,2.43)      to get  .4925
normalcdf(-1.37,2.4)   to get  .9071
normalcdf(1.37,2.4)    to get  .0778
```

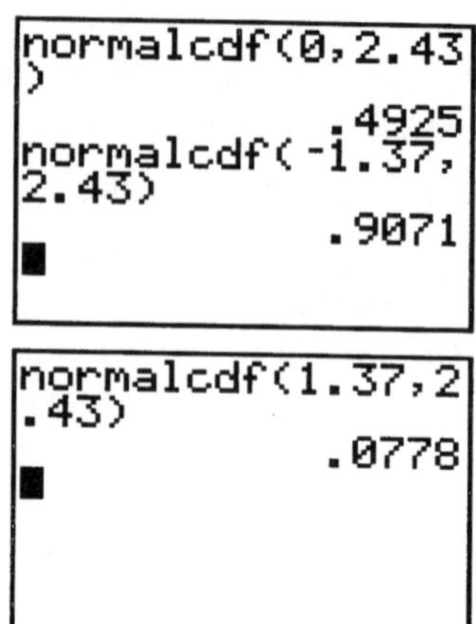

Note that the result of
`normalcdf(-1.37,2.4)` is different because
of rounding that was done in the development
of this example in the text.

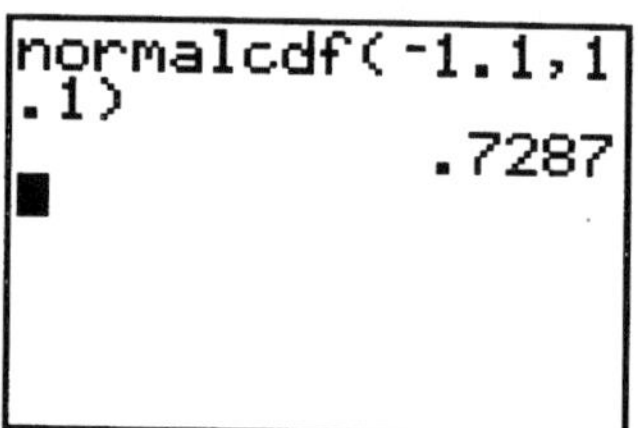

Normal Approximation to the Binomial Distribution

The work for performing a normal
approximating of a binomial distribution is done
before using the calculator. Follow the
explanation in the text of Example 4 *Coin Flips*
to see how this occurs.

Use the steps described above to evaluate
$P(-1.1 \leq Z \leq 1.1)$ using `normalcdf(-1.1,1)`. Press [ENTER] to see
the result .7287. Again, the difference in this result is due to rounding.

You're the Expert – Spotting Tax Fraud with Benford's Law

There are no new calculator skills in this section.

CHAPTER 9

MARKOV SYSTEMS

TI-83 PLUS

```
[A]*[B]→[C]
      [[700 700]]
[C]*[B]→[D]
      [[630 770]]
[D]*[B]→[E]
      [[581 819]]
```

9.1 Markov Systems – TI-83 Plus

There are no new calculator skills in this section.

9.2 Distribution Vectors and Powers of the Transition Matrix – TI-83 Plus

This section utilizes the operations of matrices that were discussed in an earlier chapter.

Distribution Vectors

This section uses repeated matrix multiplication of the initial distribution matrix [800 600] and the transition matrix $\begin{bmatrix} 0.8 & 0.2 \\ 0.1 & 0.9 \end{bmatrix}$ of Example 1 *Laundry Detergent Switching* to determine the distribution vector after a certain number of steps. Let's review how to do this.

First enter the two matrices into the calculator as matrix [A] and matrix [B].

Press [2nd] [x⁻¹] (MATRX), and move the cursor to the word EDIT. Select EDIT 1:[A]. Press [ENTER].

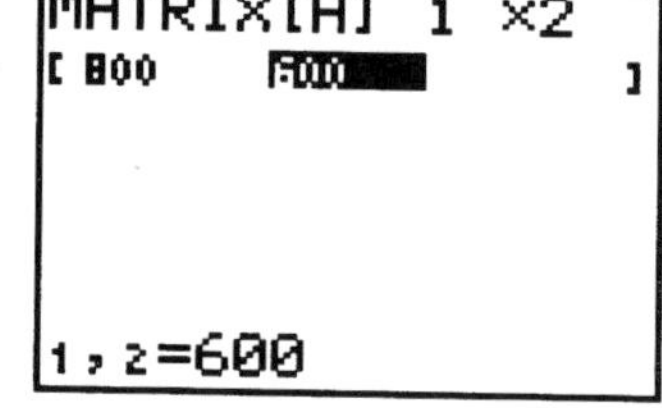

Change the dimension to 1 x 2.

Repeat these two steps selecting EDIT 2:[B] to input the 2x2 matrix [B].

Press [2nd] [MODE] (QUIT) to return to the Home screen.

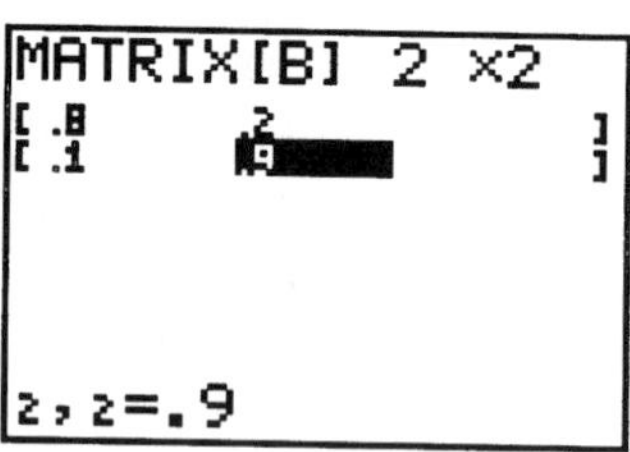

Next, multiply matrix [A] by matrix [B] to obtain the distribution after
one step.

Press 2nd x⁻¹ (MATRX), select NAMES 1:[A] to put [A] on the Home
screen. Press ×.

Press 2nd x⁻¹ (MATRX), select NAMES 2:[B] to
put [B] on the Home screen.

Store the result to matrix [C] using 2nd x⁻¹
(MATRX), select NAMES 3:[C].

Press ENTER to see the product matrix [700 700].

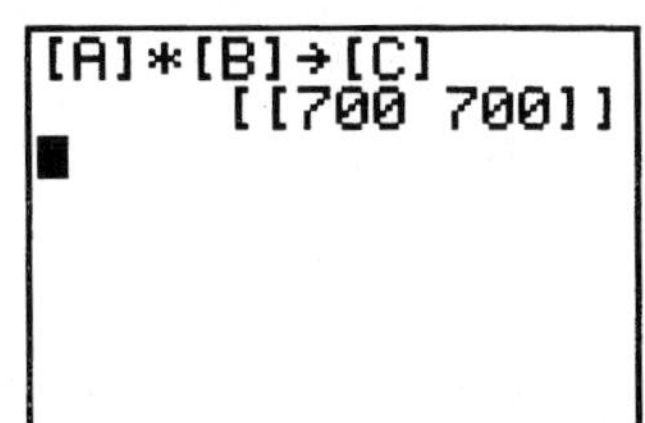

To obtain the distribution after two steps, use Matrx Names to multiply
the new matrix [C] by the transition matrix [B].

Store the result to matrix [D] using 2nd x⁻¹
(MATRX), select NAMES 4:[D].

Press ENTER to see the product matrix [630 770].

Repeat the process for as many steps as you need.

Powers of the Transition Matrix

Powers of a matrix can be found by repeated multiplication of the matrix, or by using the carat key ^.

To practice this, input matrix P of Example 2 *Investments* as matrix [A].

Press [2nd] [MODE] (QUIT) to return to the Home screen.

On the Home screen find P^2=[A]^2 and P^3=[A]^3 and compare your results with the text. Use MATRX NAMES 1:[A] to input matrix [A].

First use multiplication, then use the carat key ^. Note that you can use the [x²] key for the second power if you wish. Here are the screens you will see.

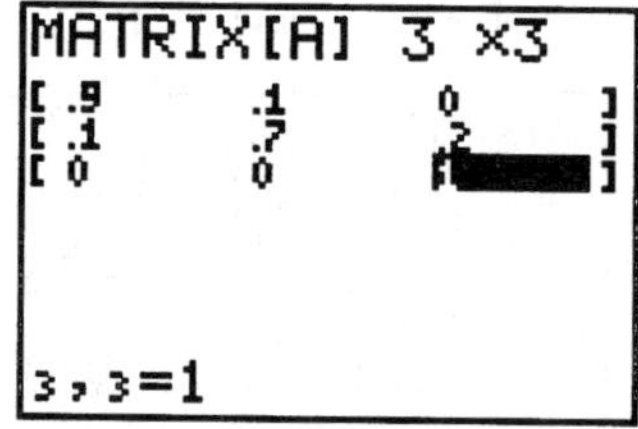

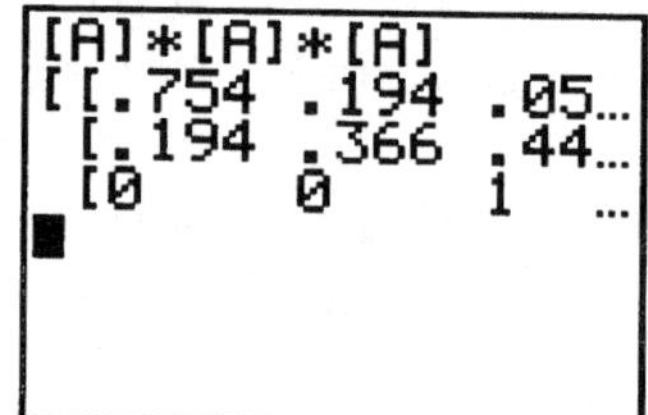

9.3 Long Range Behavior of Regular Markov Systems – TI-83 Plus

This section uses the row reducing procedures that were discussed in an earlier chapter. Let's use the augmented matrix of Example 2 *Calculating the Steady State Vector* to review these procedures.

First input the augmented matrix as matrix [A].

Press 2nd x⁻¹ (MATRX), select EDIT 1:[A]. Press ENTER.

Change the dimension to 3 x 3.

Enter the values in the matrix.

Press 2nd MODE (QUIT) to return to the Home screen.

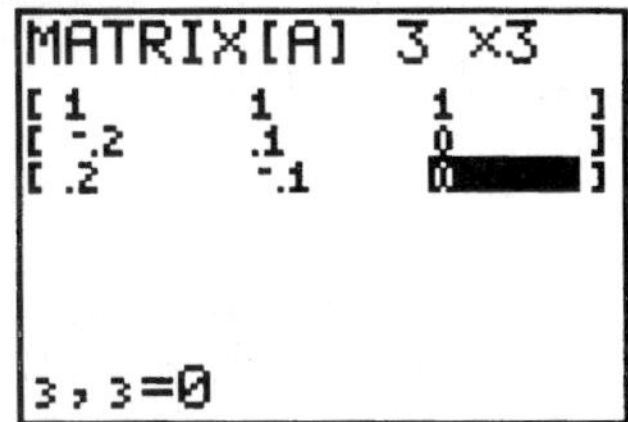

Next, clear the decimals from rows 2 and 3 use the row operation *row(in MATRX MATH.

Press 2nd x⁻¹ (MATRX), select MATH. Use the up arrow to select E:*row(Press ENTER to put *row(on the Home screen.

To multiply row 2 in matrix [A] by 10, finish the command *row(10,[A],2)→ [B]. Use MATRX NAMES to input [A] and [B].

Use the STO▸ key to store the result as matrix [B].

Repeat the line by using ⎡2nd⎤ ⎡ENTER⎤ (ENTRY).

Use the arrow keys to make the changes.

To multiply row 3 in matrix [B] by 10, finish
the command *row(10,[B],3)→ [C]. Use
MATRX NAMES to input [B] and [C].

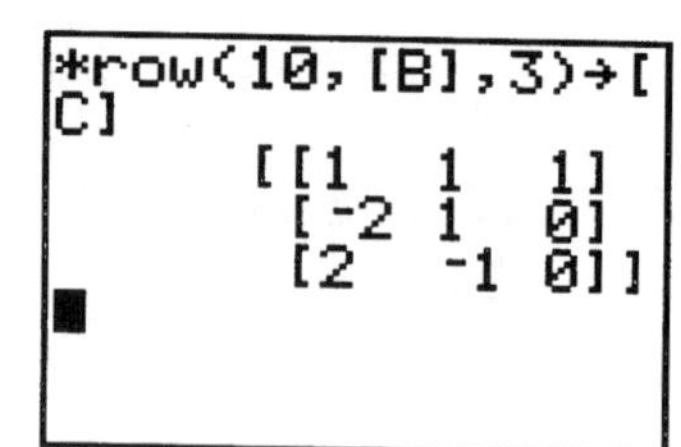

Use the ⎡STO▸⎤ key to store the result as matrix [C].

Next, change the remaining entries of column 1 to $\emptyset$'s.
To change a_{12} to a $\emptyset$, perform $R_2 + 2 R_1$ use the row operation *row+(
in MATRX MATH.

Press ⎡2nd⎤ ⎡x^{-1}⎤ (MATRX), select MATH. Select
F:*row(. Press ⎡ENTER⎤ to put *row+(on the
Home screen.

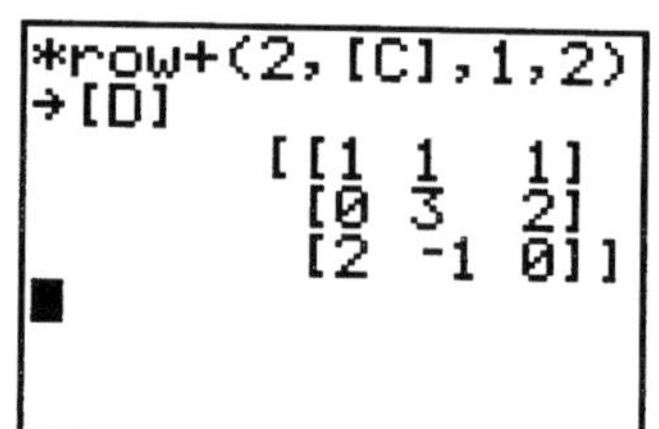

Finish the command *row+(2,[C],1,2)→ [D].
Use MATRX NAMES to input [C] and [D].

Use the ⎡STO▸⎤ key to store the result as matrix [D].

To change a_{13} to a $\emptyset$, perform $R_3 - 2 R_1$ use the row operation *row+(
in MATRX MATH.

Press ⎡2nd⎤ ⎡x^{-1}⎤ (MATRX), select MATH. Select
F:*+row(. Press ⎡ENTER⎤ to put *row+(on the
Home screen.

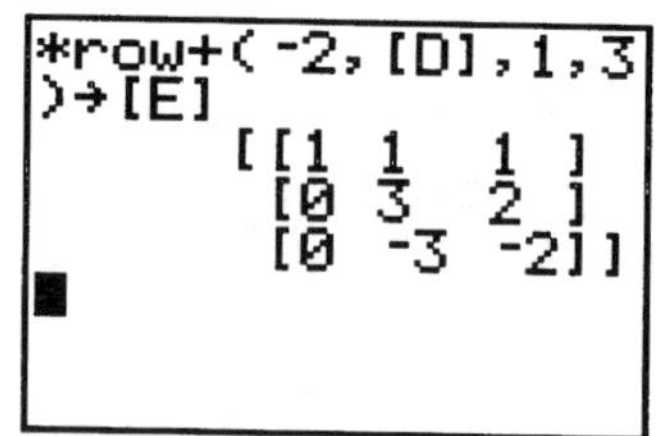

Finish the command *row+(-2,[D],1,3)→
[E]. Use MATRX NAMES to input [D] and
[E].

Use the ⎡STO▸⎤ key to store the result as matrix [E].

To change a_{21} to a 0, perform $3R_1 - R_2$ using two steps. First multiply row 1 by 3, then use *row+(multiply row 2 by −1, add the result to row 1 and store the result in row 1.

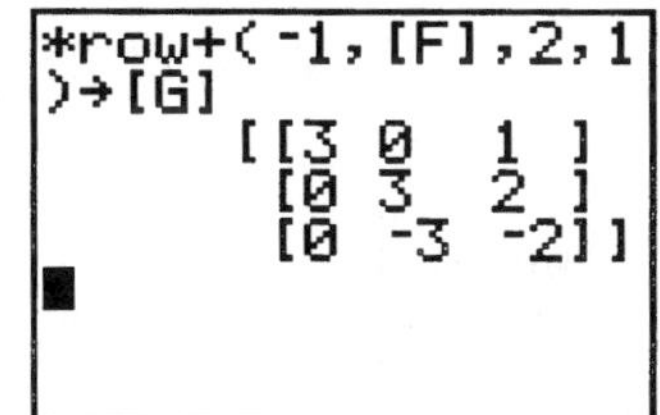

To change a_{23} to a 0, perform $R_3 + R_2$ using row+(.

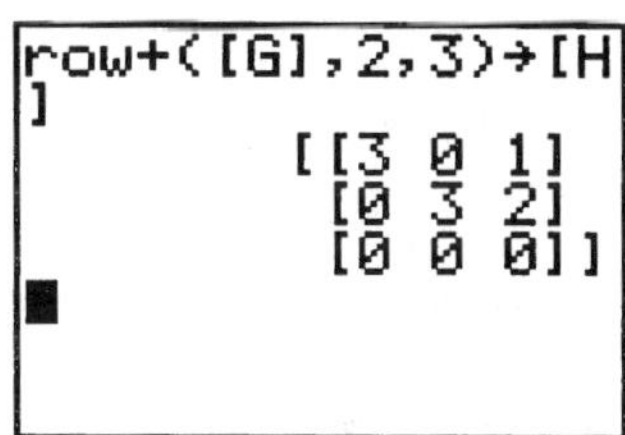

Lastly, multiply rows 1 and 2 by 1/3 using *row(.

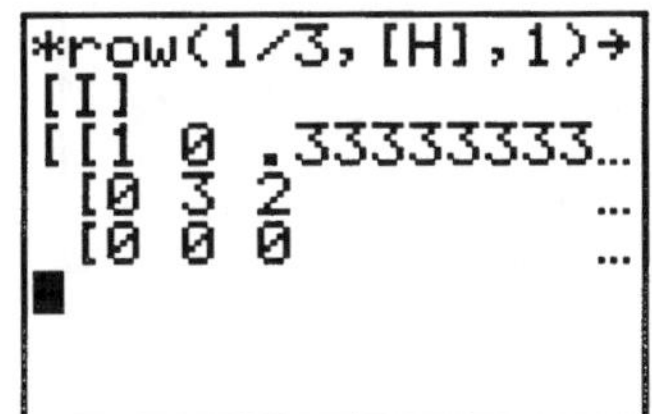

Finally, we can convert the resulting matrix [J] to fractions using
MATH MATH 1:Frac.

9.4 Absorbing Markov Systems – TI-83 Plus

This section uses the identity matrix and matrix inverses as discussed in earlier chapters. Let's use Example 1 *Profitability* to review these skills.

In Example 1 *Profitability* we see that matrix S is $\begin{bmatrix} 0.9 & 0.1 \\ 0.1 & 0.7 \end{bmatrix}$. We must compute $Q = (I - S)^{-1}$.

First input matrix S as matrix [A].

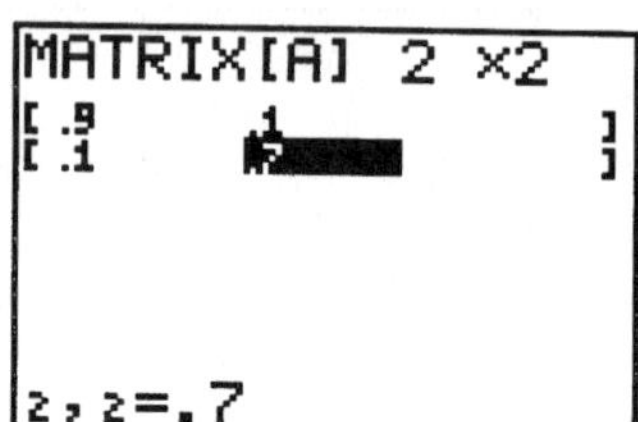

Return to the Home screen.

Next create a 2 x 2 identity matrix using MATRX MATH 5:identity and store it as matrix [B].

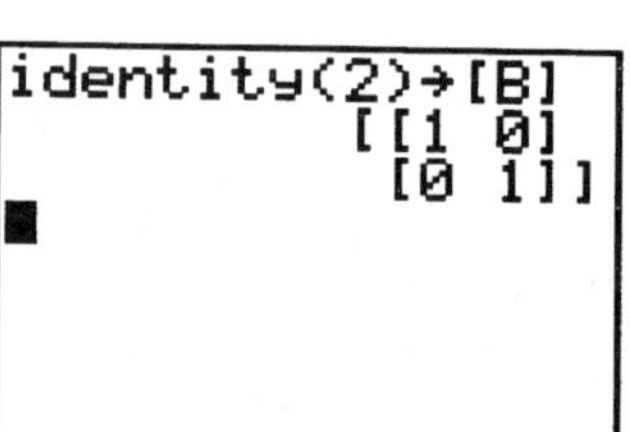

On the Home screen input [B]-[A] using MATRX NAMES and store it to matrix [C].

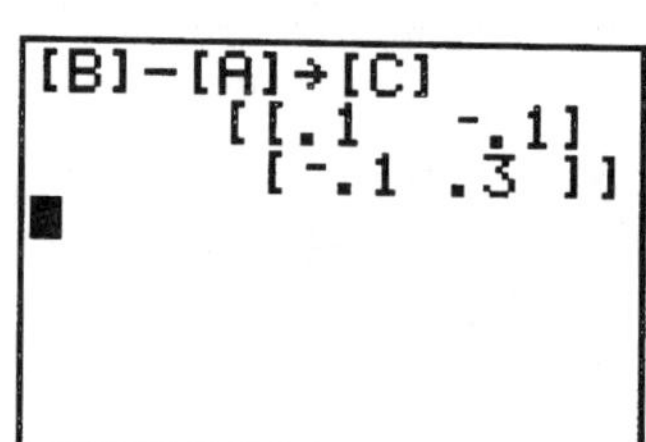

Lastly, find the inverse of matrix [C] using $\boxed{x^{-1}}$.

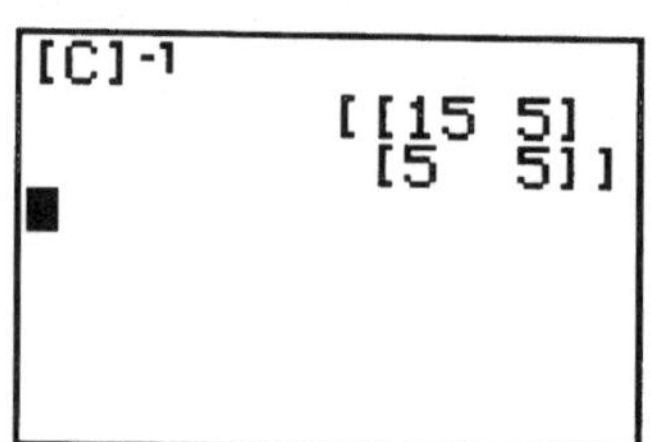

CHAPTER 1

LINEAR FUNCTIONS AND MODELS

TI-86

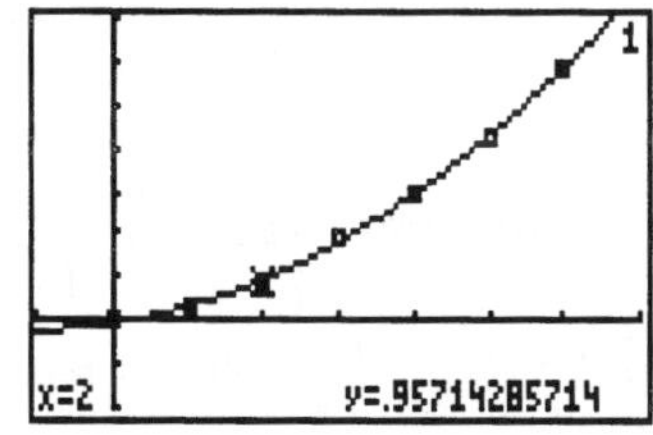

1.1 Functions from the Numerical and Algebraic Viewpoints

In this section you will learn several different ways of using the TI-86 to evaluate functions. As a result, some of the features of the calculator will be illustrated, including the equation editor, memory storage, TABLE, and lists. The TRACE feature in the graph editor will also be used.

Evaluating Functions

Evaluating functions is a simple yet important mathematical skill. In section 1.1 Example 3, you were shown how to evaluate the function $f(x) = 0.03x^2 - 0.06x + 6$ for $x = 0, 1, 2, \ldots, 10$. We will use this same function to show how the TI-86 can be used to evaluate this function. Please note that prior to beginning each method, the home screen is cleared.

Method 1 – Using y1 on the Home Screen

The first method of evaluating this function using the TI-86 is to use y1 from the equation editor on the home screen. Press [GRAPH] and then press [F1] (y (x)=) to access the equation editor. With the cursor at \y1=, enter $.03x^2 - .06x + 6$ by pressing .03 [x-VAR] [x²] [−] .06 [x-VAR] [+] 6.

Return to the home screen by pressing [2nd] [QUIT]. On the home screen, press [2nd] [alpha] [y] 1 [(] 0 [)] [ENTER] to evaluate f at $x = 0$. The calculator should return the value 6.

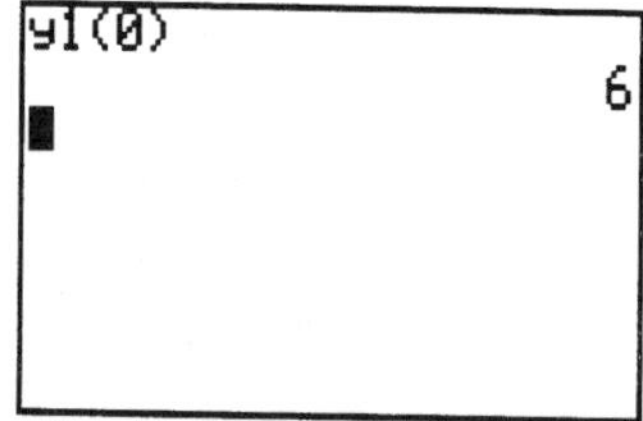

Repeat the previous line on the home screen by pressing [2nd] [ENTER]. Use the left arrow to position the cursor over the 0, press 1 to overwrite it, and then press [ENTER]. The calculator returns the value 5.97, showing that $f(1) = 5.97$. Continue in this same manner for $x = 2, \ldots, 10$ to find $f(2), \ldots, f(10)$.

120

Method 2 – Using y1 with x-values Stored in Memory

This method involves using y1 on the home screen with values stored in memory. To begin, clear the home screen by pressing [CLEAR].

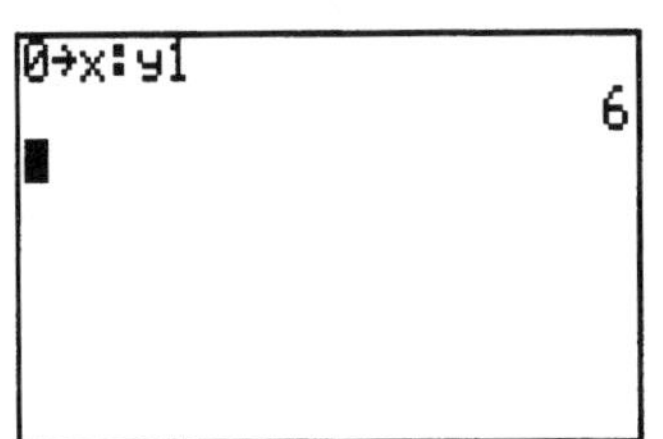

Press 0 [STO▸] [X-VAR] [2nd] [:] [ALPHA] (the colon allows you to enter multiple expressions or instructions on one command line). Now press [2nd] [alpha] [y] [1] [ENTER]. The value 6 will appear on the screen. The expression 0➔x is the calculator's syntax for "let x equal 0". The variable x has now assumed the value 0 and has been stored in memory. The variable x will retain its value of 0 until changed.

Remember, the function $f(x) = .03x^2 - .06x + 6$ has been defined in the calculators equation editor as y1. The line you see on the screen is basically communicating "when x is 0, find what y is". The calculator evaluates the function for $x = 0$ and returns a value of 6 for y.

Repeat the previous line on the home screen by pressing [2nd] [ENTER]. Use the left arrow to position the cursor over 0, press 1 to overwrite it, and then press [ENTER]. The calculator returns the value 5.97. You can now repeat this process for $x = 2, \ldots , 9$.

You must be careful when changing the value of x from 9 to 10. After evaluating the function for $x = 9$, press [2nd] [ENTER] to recall the last command. After positioning the cursor over the 9, press 1 [2nd] [INS] 0 to replace the 9 with 10. Pressing the 0 without using the insert editing feature INS will overwrite the ➔ symbol needed to store the new value.

Method 3 – Using y1 with a List of Values

y1 can also be used on the home screen to evaluate f for a list of values. Noting that $x = 0, 1, 2, \ldots , 10$ represents a list of values, we can evaluate $f(x)$

for this list, with the values of f being returned in a list. One way to do this is the following.

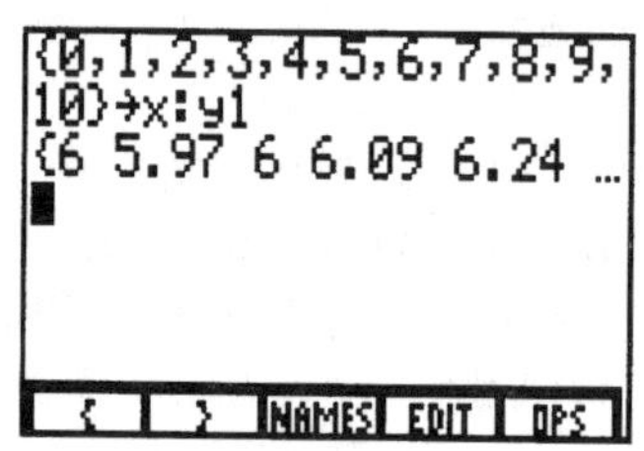

On a clear home screen, press 2nd [LIST] to access the LIST menu. Press F1 to select the open brace symbol { and to paste it on the screen. Press 0 , 1 , 2 , 3 , 4 , 5 , 6 , 7 , 8 , 9 , 10 F2 (to select the close brace }) STO▸ x-VAR 2nd [:] ALPHA. This will store the list of values as x. Now press 2nd [alpha] [y] 1 ENTER. The calculator will return the list of values representing $f(x)$ for each x-value in the stored list.

Note the ellipsis (...) at the end of the list. These indicate that there is more to the list than what is seen on the screen. Use ▶ to view the rest of the list. The end of the list is shown in the screen at the right. You can use ◀ to scroll back to the beginning of the list.

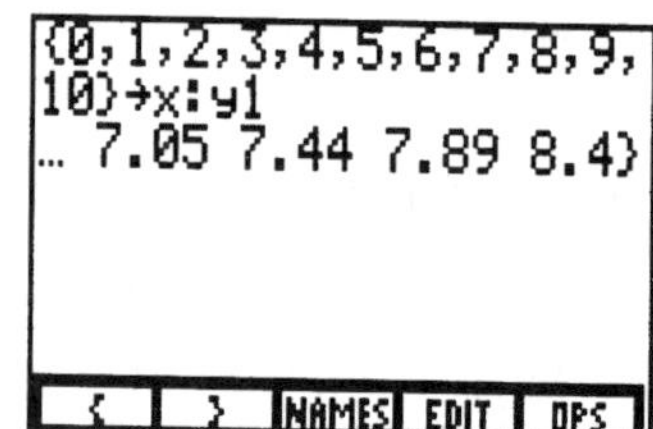

Method 4 – Using the TABLE Feature

We can also evaluate f using a table. The table feature of the TI-86 is a wonderful tool for obtaining the values of a function at several values of x.

Access the equation editor (GRAPH F1) and make sure that $f(x)$ is defined as y1. If not, then at y1 type $.03x^2 - .06x + 6$. If you have been following along from the start of this section, the function should be defined already.

Now press TABLE F2 (TBLST) to access the TABLE SETUP editor. Set TblStart to 0 and press ENTER. This tells the calculator at what value of x to start the table for viewing. Set ΔTbl to 1 and press ENTER. This indicates to the calculator to increase x by 1 throughout the table. Set the Indpnt: option to Auto by pressing ENTER. Auto should be highlighted.

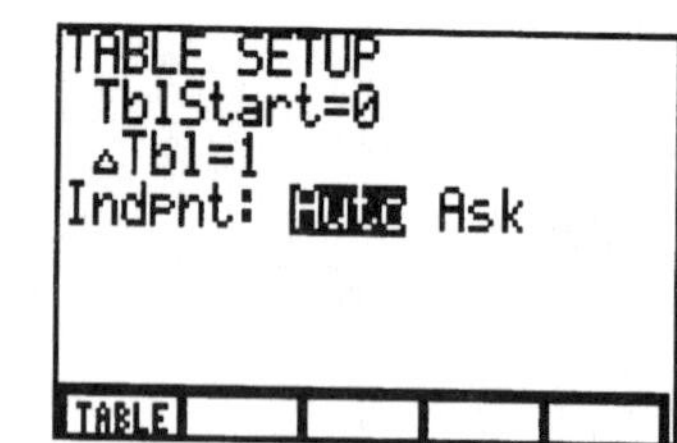

You are now ready to view the table. Press F1 (TABLE) to view the table. You can press ▾ several times to scroll down the table to view the value of f

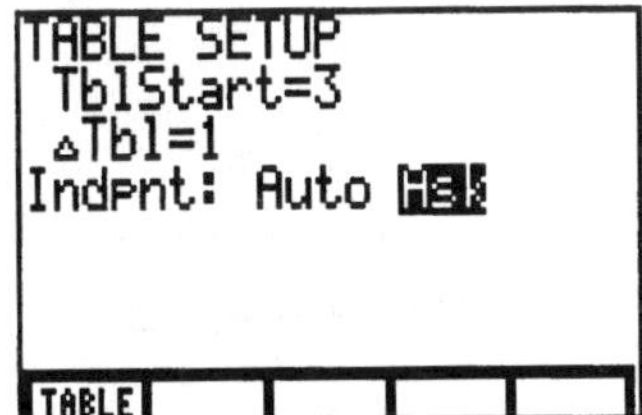

(given in column 2) for different integer values of x. You can scroll back up the table by pressing ▲ several times.

An alternate way of using the table feature of the TI-86 is to obtain function values for specific values of x that you enter yourself. This is opposed to having the calculator generate several values at once automatically as illustrated above.

Press [F1] (TBLST) to access the TABLE SETUP editor. Using this method, the

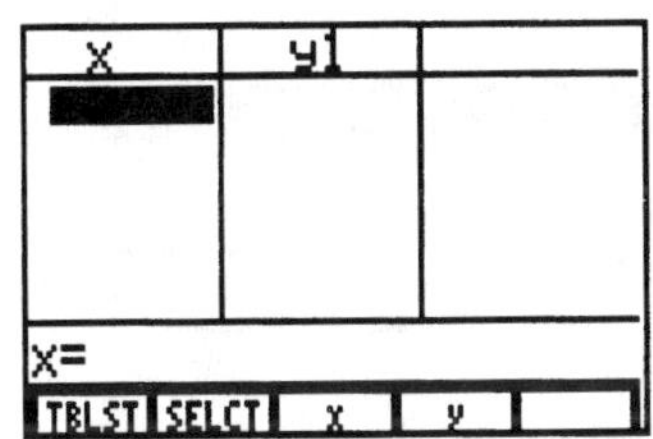

values of TblStart and ΔTbl are irrelevant, so these values on your screen may be different from those shown on my screen, and they need not be changed. Press ▼ ▼ ▶ [ENTER] to set the Indpnt: option to Ask. This prompts the calculator to allow you to input values for the independent variable x.

Now press [F1] (TABLE) to access the table. Note that no values are shown for x nor y. This is because the calculator is "asking" you to input a value for x and it will then compute the corresponding value of y.

Input 0 and press [ENTER]. The number 0 will appear in the first column and the value of the function at x = 0, i.e., $f(0)$, will appear in the second column. Thus, $f(0) = 6$. Now input 1 and press [ENTER].

Try to continue for $x = 2, 3, \ldots , 10$, pressing [ENTER] after inputting each value. You will realize that there is a limitation to the Ask option. This option only allows you to enter up to six values. Attempting to input 6 will only overwrite the number 5 and return the value of f at 6 in column 2. Input the remaining values for x and observe what happens.

Method 5 – Using the Graph

The final method to be illustrated for evaluating a function is from the graph of the function on the graph screen. Access the equation editor (remember, press GRAPH F1). If not done yet, define the function in question as y1 (in this case, \y1=.03x^2-.06x+6).

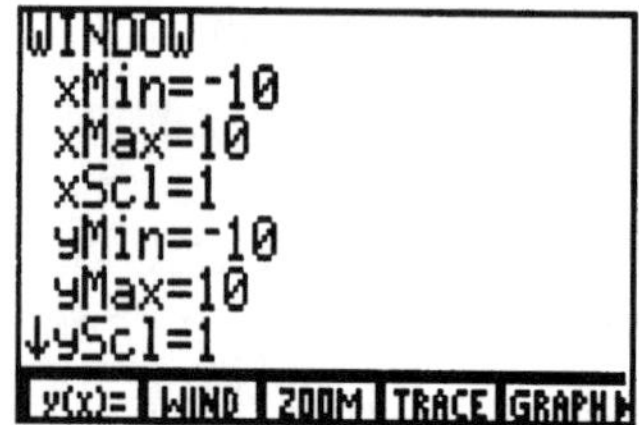

Press EXIT F2 (WIND) to access the window editor. Choose the standard viewing window settings: xMin=-10, xMax=10, xScl=1, yMin=-10, yMax=10, yScl=1. These window settings are basically chosen by trial and error.

Once the window settings are in place, press F5 (GRAPH) to view the graph of $f(x)$. (If you are later working with another function and the graph does not appear on the screen, use another window setting.)

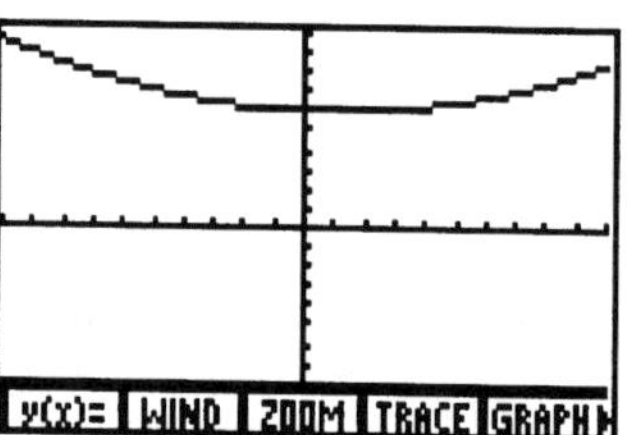

Press F4 (TRACE) to activate the trace cursor. The numbers at the bottom of the screen show the value of x and the corresponding value $f(x)$ (or y). Thus, the screen shows that when $x = 0$, the value of the function is $y = f(0) = 6$.

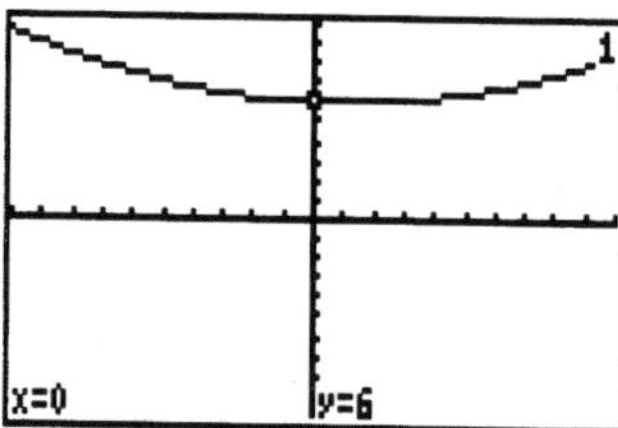

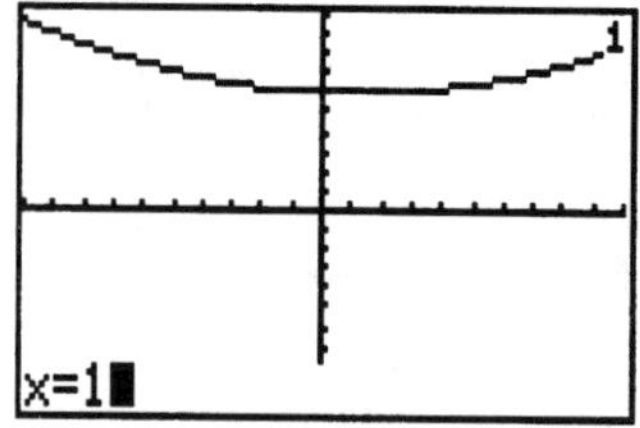

With TRACE still activated, press the number 1. The bottom of the screen will read x=1. Press ENTER. The bottom of the screen will show x=1, y=5.97, showing that $f(1) = 5.97$. Also, the cursor jumps to the corresponding spot on the graph. You can now enter the other values of x at which you wish to evaluate f.

Testing an Inequality

The TI-86's TEST feature will allow you to test whether an inequality is true or false. Before starting, however, press EXIT as often as needed to return to the home screen. Press CLEAR to clear any data shown on the home screen.

The TEST menu can be accessed by pressing 2nd [TEST]. This is where the inequality symbols can be found. These symbols can be used to test whether an inequality is true or false.

For example, if we were to test whether 2 is less than or equal to 10, we would enter 2 2nd [TEST] F4 ($\leq$) 10 ENTER. The calculator will return a value of 1, indicating the inequality is true. To test the statement "2 is greater than 10", we press 2 F3 (>) 10 ENTER and observe the value 0 appear, indicating that the statement is false.

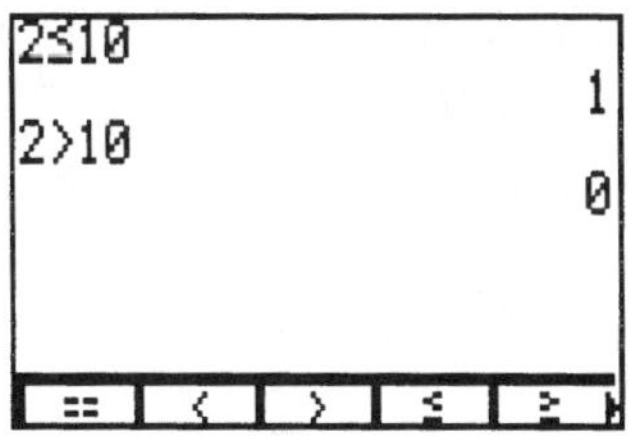

1.2 Functions from the Graphical Viewpoint

In this section you will learn how to set up the STAT (statistics) editor to include lists L1, L2, ..., L6. You will also learn how to plot individual points on the graph screen, input piecewise functions for graphing, and graph a variety of functions.

Setting up the STAT Editor

In order to plot individual points on the graph screen, we will need to use the TI-86's STAT editor (also called the LIST editor). Before we begin however, we will set up the STAT editor to actually look somewhat like the TI-83's STAT editor. This will make accessing lists defined in the editor somewhat easier. It will also increase the number of lists in the editor that are readily available for use.

Press [2nd] [STAT] to view the STAT menu. 

Enter the STAT editor by pressing [F2] (EDIT). There will be three columns shown for you to use for lists of values.

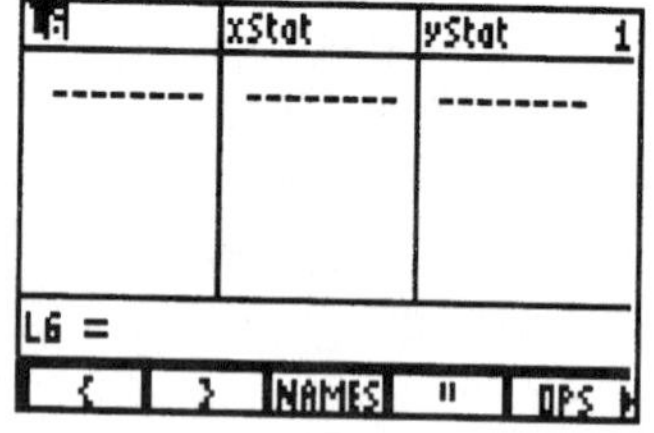

Arrow to the very top of the column titled xStat. At the top of this column, press [2nd] [INS] to insert a column. A new column will be created. The calculator is now prompting you to give the new column a name (as indicated by the Name= prompt at the bottom of the screen, above the menu line). The calculator is also in ALPHA mode (as indicated by the blinking ⓐ next to the Name= prompt). Press [L] [ALPHA] 6 [ENTER] to name the column L6.

The cursor should still be located at the top of column L6. While there, create another column by pressing [2nd] [INS]. To name this new column, repeat the

126

instructions given in the above paragraph, last sentence, replacing the "6" by a "5". The screen for this is shown with the next paragraph.

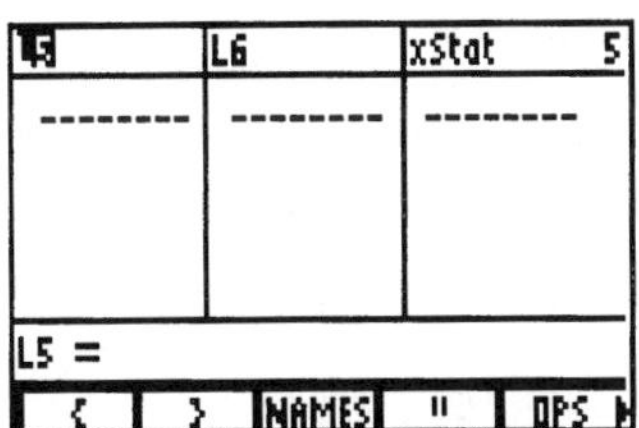

Continue to create columns L4, L3, L2, and L1, replacing the "5" with the numbers 4, 3, 2, and 1, respectively, each time. Do this slowly and carefully, as it is easy to make a mistake if you attempt to complete this task to quickly. (Note that you are creating the columns in descending order. This will keep the column reference number in the upper right corner consistent with the name of the list.)

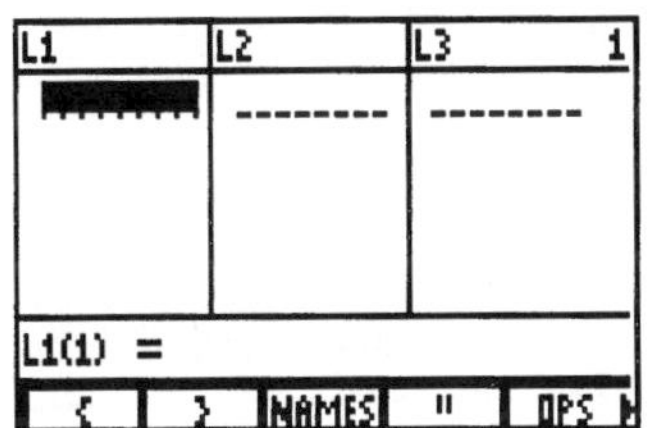

When you have completed creating and naming column L1, press ▼ and use ▶ and ◀ to view the new setup of your STAT editor!

Plotting Individual Points on the Graph Screen

Example 2 in the text asks you to plot the following points for
$R(p) = -5,600p^2 + 14,000p$ $(0 \le p \le 2.5)$.

p	0	0.5	1	1.5	2	2.5
$R(p) = -5,600p^2 + 14,000p$	0	5,600	8,400	8,400	5,600	0

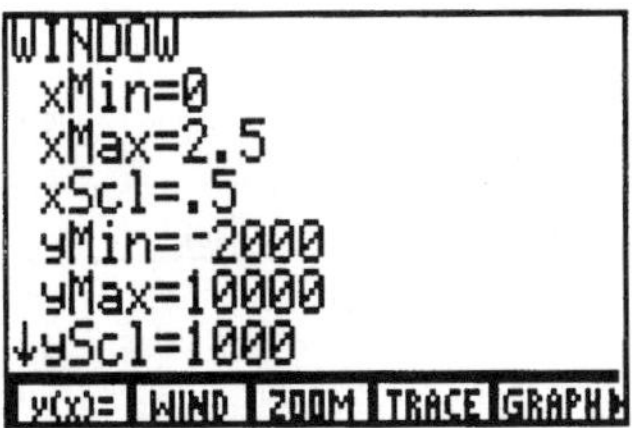

Starting at the home screen, press GRAPH F2 (WIND), and set the window to xMin=0, xMax=2.5, xScl=.5, yMin=-2000, yMax=10000, yScl=1000. (We set yMin to −2000 so that the p-axis is not covered by the menu line after the graph is displayed.)

Access the equation editor (GRAPH F1) and press F5 (SELCT) to deselect (or press F4 (DELF) to delete) any functions that are selected for graphing. A deselected equation is indicated by the equal sign no longer being highlighted. Press F5 (SELCT) several

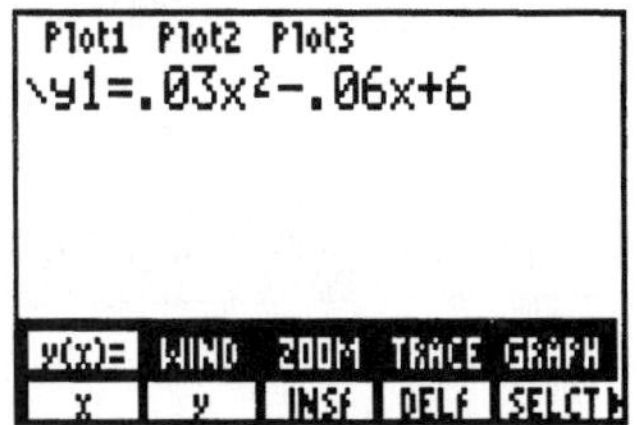

times to see the difference between a selected function (equal sign highlighted) and a deselected function (equal sign not highlighted).

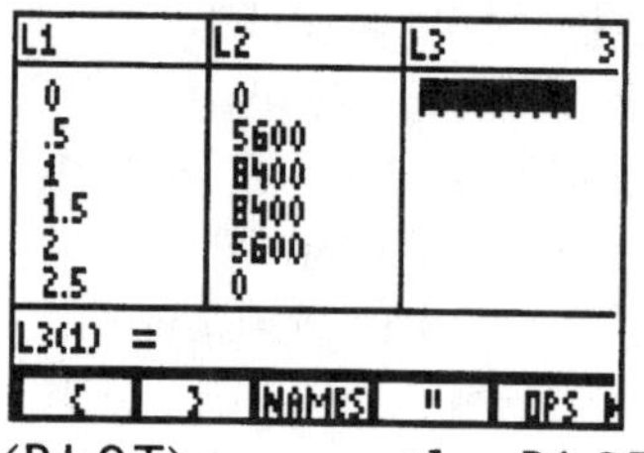

To enter the points, press [2nd] [STAT] [F2] (EDIT) to get into the STAT editor. Enter the points, using L1 for the p values and L2 for the $R(p)$ values. Once these values are entered, press [2nd] [STAT] [F3] (PLOT) access the PLOTS editor.

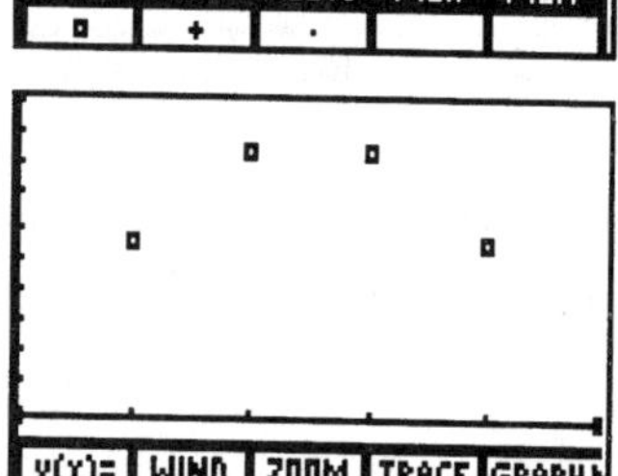

In the PLOTS editor press [F1] (PLOT1) to select plot 1. Then do the following.

1. Press [ENTER] to turn on the plot.
2. Press [▼] and [F1] (SCAT) to select scatter plot.
3. Press [▼] and then [F4] to choose L1 for Xlist Name=.
4. Press [▼] and then [F5] to choose L2 for Ylist Name=.
5. Press [▼] [F1] for selecting the box as the Mark type.
6. Press [GRAPH] [F5] (GRAPH) to see the plot of the points.

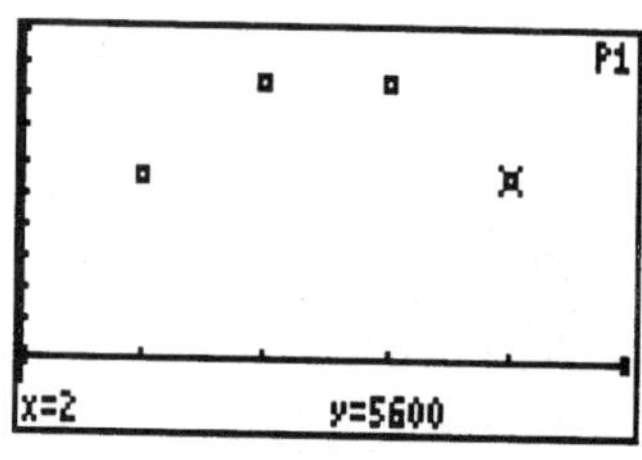

When the plot is complete, activate the trace cursor by pressing [F4] (TRACE), and use [▶] and [◀] to move from point to point. The x and y coordinates of each point are displayed at the bottom of the screen, and P1 (for PLOT1) is displayed in the upper right corner, identifying the plot.

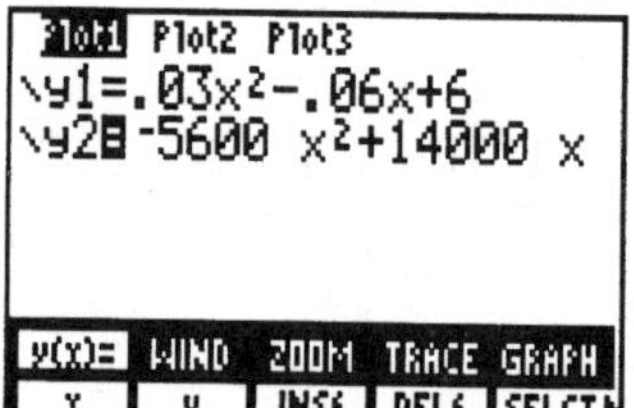

If you enter $R(p) = -5{,}600p^2 + 14{,}000p$ in the equation editor as y2 (using X in place of p) and graph ([EXIT] [F5]), you will see the graph of the equation going through the points.

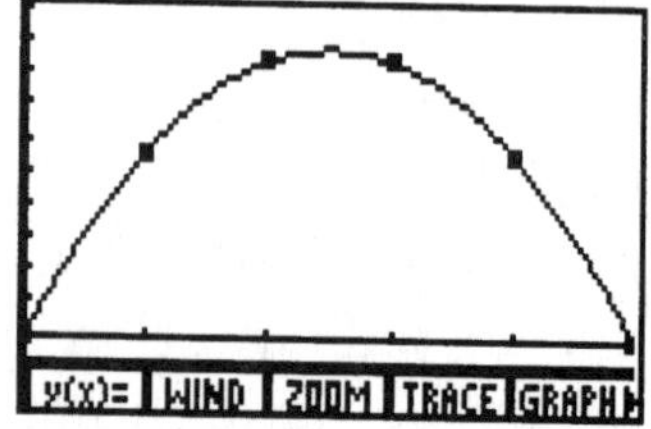

Inputting Piecewise Functions for Graphing

Inputting piecewise defined functions to be graphed on the TI-86 can be a bit
tricky. However, with a bit of practice, you can learn how to obtain graphs of
piecewise defined functions with your calculator. You will need to use the
inequality symbols, which are found in the [TEST] menu over $\boxed{2}$. As an
example of how to use these symbols to input a piecewise function into the
TI-86, we can graph the function

$$U(t) = \begin{cases} 76 - 3.3t & if \ \ 1 \leq t \leq 10 \\ 18 + 2.5t & if \ \ 10 < t \leq 14 \end{cases}$$

from Example 3 of section 1.2 in the text.

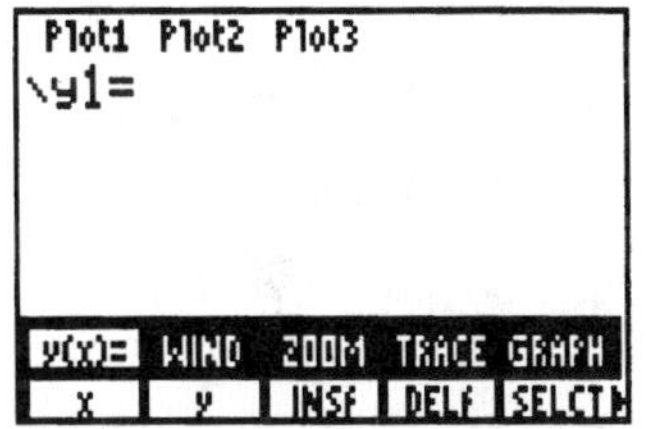

Press $\boxed{\text{GRAPH}}$ $\boxed{\text{F1}}$ to access the equation editor. If
necessary, press $\boxed{\text{F4}}$ (DELF) as often as needed to
delete all unwanted or unneeded functions that are
currently defined in the function editor. It may also
be necessary for you to turn off any plots that are on
(highlighted) by pressing the up-arrow key and $\boxed{\text{ENTER}}$.

With the cursor at \y1=, press $\boxed{\text{MORE}}$ and then $\boxed{\text{F3}}$ (STYLE) six times to select
dot mode as the graph style. Now enter the function U as $\cdot.$y1=(76-
3.3x)(1≤x)(x≤10)+(18+2.5x)(10<x)(x≤14). We must enter $1 \leq x \leq 10$
as $(1 \leq x)(x \leq 10)$ because the calculator will not accept the syntax $1 \leq x \leq 10$.
The function is shown over the three screens given below.

 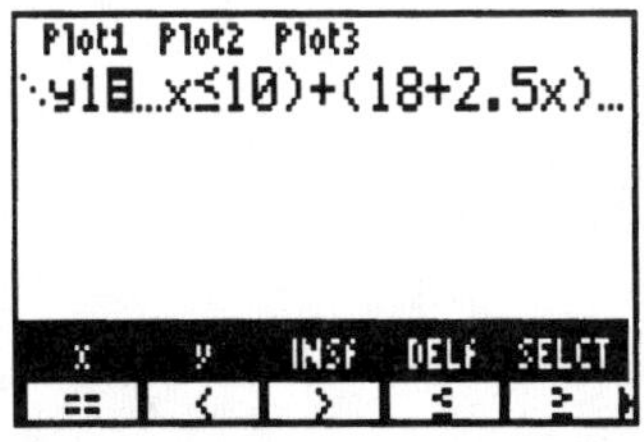 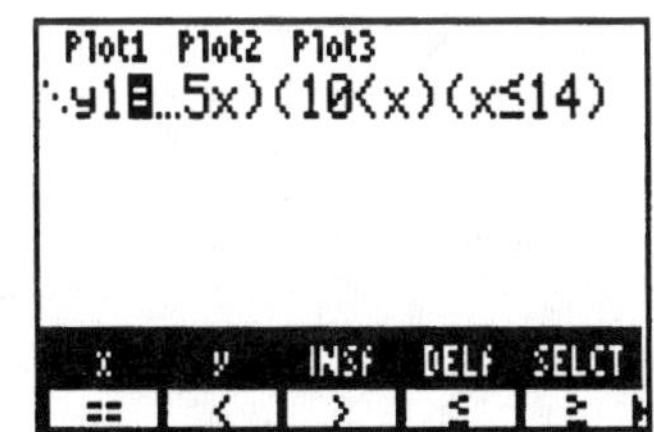

Adjust the window ($\boxed{\text{GRAPH}}$ $\boxed{\text{F2}}$) to xMin=-1, xMax=15,
xScl=1, yMin=-20, yMax=80, yScl=10, and then
press $\boxed{\text{F5}}$ (GRAPH) to see the graph.

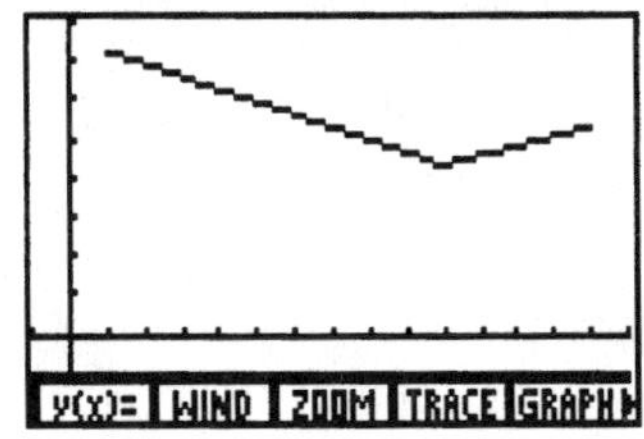

Inputting a Variety of Functions into the Equation Editor

One must be extremely careful how he or she enters various functions for graphing. The correct use of parentheses, the negation key $\boxed{(-)}$, the TI-86 built-in functions, and order of (algebraic) operations is very important in regards to how the calculator will interpret the expressions you define.

Below are some examples of functions that can cause problems if not entered correctly into the equation editor. The TI-86 window screen shows how these functions should be entered. The location of some of the functions needed, in particular the absolute value command $\mathtt{abs}$ and the nth root symbol $\sqrt[x]{}$, will be discussed in the next section.

1. $f(x) = \sqrt{x+2}$

2. $f(x) = \sqrt[3]{x} - 1$

3. $f(x) = x^{2/5}$

4. $f(x) = \dfrac{x+2}{x-1}$

5. $f(x) = 2 - |\,x - 1$

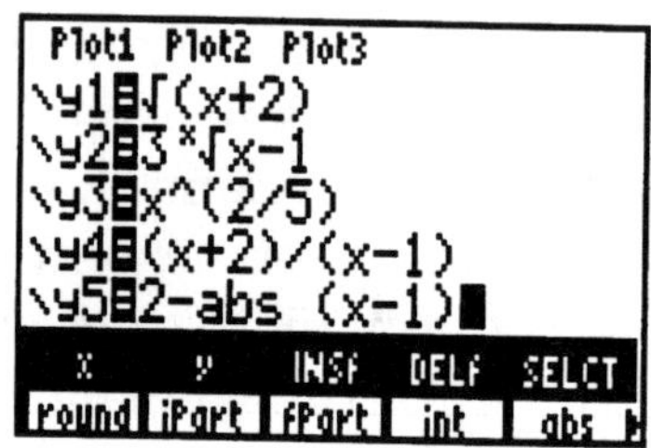

1.3 Linear Functions

In this section you will learn how to create your own CUSTOM menu, express an answer as a fraction, and how to fit a linear equation to data.

Customizing Your Own Menu

From a clear home screen, press [CUSTOM] to display your custom menu. If your calculator is new, your custom menu should be empty.

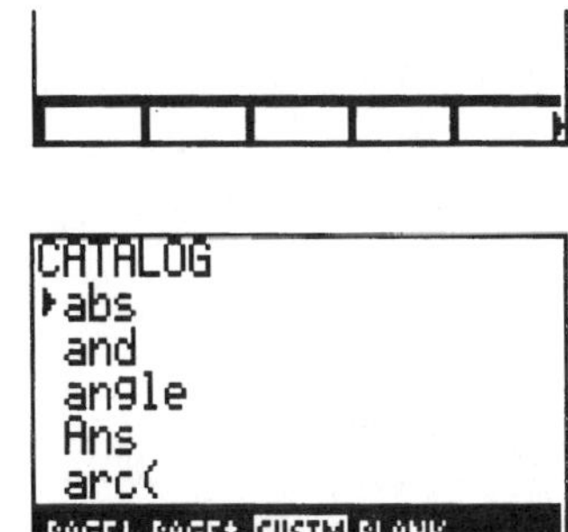

Press [2nd] [CATLG-VARS] [F1] (CATLG) to access the calculator's catalog of all TI-86 built-in functions in alphabetical order. Press [F3] (CUSTM) to select the custom option in the calculator's catalog. The ALPHA-lock is now on, so press [A] ([LOG]) to take you to the top of the catalog listing.

With the selection cursor (▶) pointing at the item abs, press [F1] and the selection will appear in the first menu location. This is the absolute value function. Press [▲] until the selection cursor is at the ▶Frac command; press [F2]. Press [▲] several times again until the selection cursor is at the symbol $\sqrt[x]{}$ and press [F3].

Now press [2nd] [QUIT] to exit the catalog and return to the home screen. Press [CUSTOM] to view your custom menu.

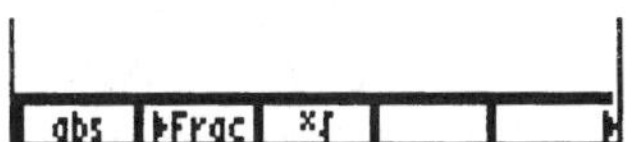

Expressing an Answer as a Fraction

As part of Example 4 part b of this section of the textbook, you needed to find the slope of the straight line through (1, 2) and (3, −1). You can do this on the home screen by entering (-1-2)/(3-1) and pressing [ENTER] to obtain the result −1.5. Be sure to use the subtraction sign [−] between terms and the

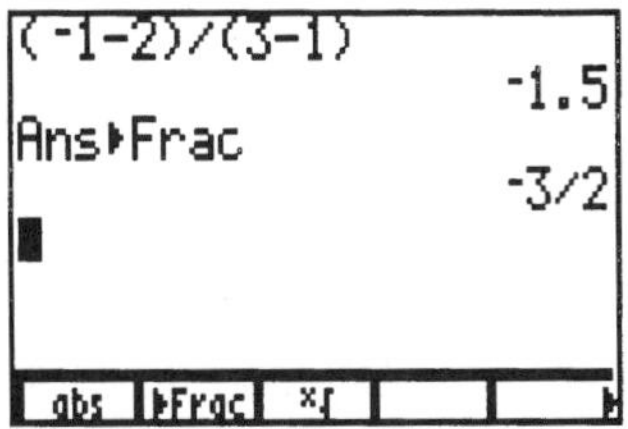

131

negation sign $\boxed{(-)}$ for negative numbers such as -1. Now press $\boxed{\text{CUSTOM}}$ $\boxed{\text{F2}}$ ($\blacktriangleright$Frac) $\boxed{\text{ENTER}}$ to convert the last answer to a fraction.

Fitting a Linear Equation to Data

Also in Example 4 part b, you were asked to find the equation of the straight line through the points $(1, 2)$ and $(3, -1)$. To use the TI-86 to fit a line to these data points, access the STAT editor ($\boxed{\text{2nd}}$ [STAT] $\boxed{\text{F2}}$). Clear any existing data from lists L1 and L2. Do this by arrowing up to the very top of L1, press $\boxed{\text{CLEAR}}$ and then $\boxed{\text{ENTER}}$. List L1 will be cleared. Do the same for list L2.

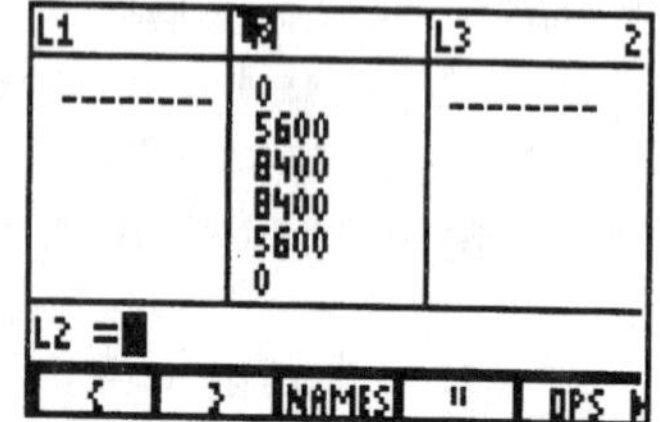

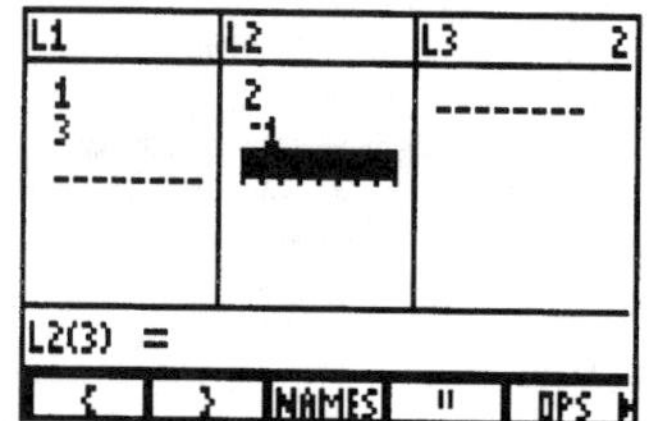

Enter the x coordinates into L1 and the corresponding y coordinates into L2, and then exit to the home screen. Now press $\boxed{\text{2nd}}$ [STAT] $\boxed{\text{F1}}$ (CALC) $\boxed{\text{F3}}$ (LinR) $\boxed{\text{ALPHA}}$ [L] 1 $\boxed{,}$ $\boxed{\text{ALPHA}}$ [L] 2. This command is used to find the linear equation that best fits the data points listed in L1 and L2.

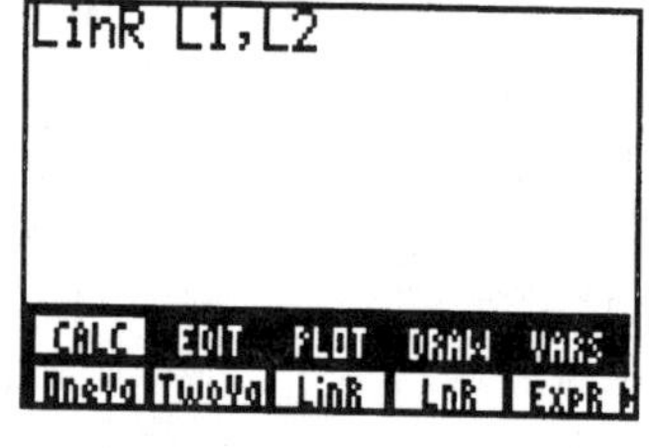

Press $\boxed{\text{ENTER}}$. The calculator displays the equation of the line as $y = a + bx$ and not $y = ax + b$, so you must pay attention to where the values of a and b should be placed. Thus, the equation of the line through the two points is shown to be $y = -1.5x + 3.5$ in slope-intercept form. Note that this equation is equivalent to the one found in the text.

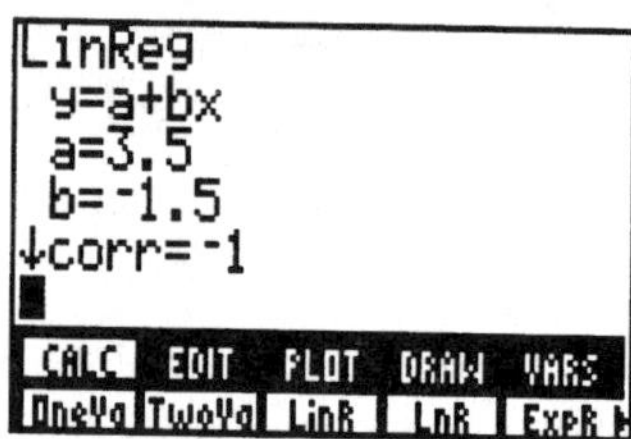

1.4   Linear Models

There are no new calculator skills in this section.

1.5   Linear Regression

This section discusses how to find the correlation coefficient. For the TI-86, when carrying out the procedure discussed in section 1.3 of this manual (see the portion titled *Fitting a Linear Equation to Data*), the value of the correlation coefficient r is displayed after `corr=`.

You're the Expert – Modeling Spending on Internet Advertising

In this section, a quadratic model for a data set was found together with its correlation coefficient. Also, the graph of this model together with the data was produced. Unfortunately, although the TI-86 will compute and display the correlation coefficient for linear models, it does not do so for other models. Thus, the correlation coefficient for a quadratic model can not be found directly, so assessing the goodness of fit of the model to the data can only be done visually (with the graph) and not numerically. Therefore, you will find a quadratic model and graph it together with the data points.

If you have not already done so, read section 1.2 – *Functions from the Graphical Viewpoint* on how to set up your STAT editor. This will be needed (and will make it easier) to follow along with the discussion below.

Finding a Quadratic Model

To find a quadratic model for the advertising spending example, first access the STAT editor (2nd [STAT] F2) and clear lists L1 and L2 of all data.

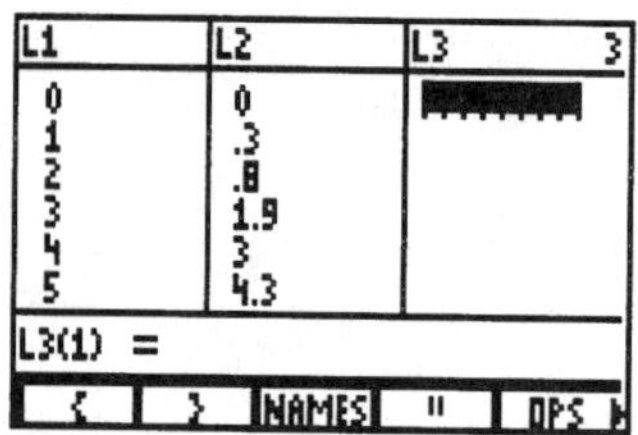

Input the data into L1 and L2 in the STAT editor. Since $x =$ years since 1995, the value $x = 0$ represents 1995, $x = 1$ represents 1996, etc. After entering all the data into the list, exit to the home screen.

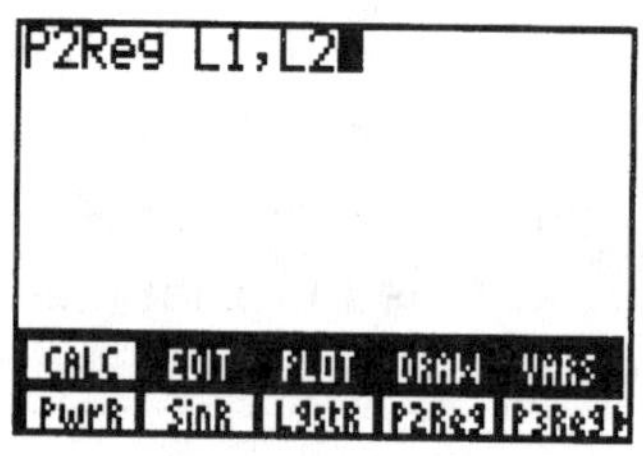

Press [2nd] [STAT] [F1] (CALC) [MORE] [F4] (P2Reg) [ALPHA] [L] 1 [,] [ALPHA] [L] 2. This is the quadratic regression feature. Press [ENTER]. The coefficients a, b, and c of the quadratic model is returned in a list called PRegC. You can use the left and right arrow keys to view the entire list.

Inputting the Regression Equation into the Equation Editor

You can input the regression equation into the equation editor at the same time that you find the model.

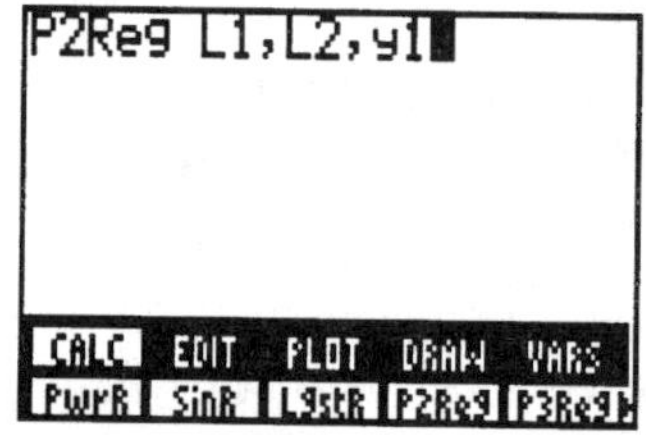

To do this, clear the home screen with [CLEAR]. Press [2nd] [ENTER] to recall the last entry (P2Reg L1,L2). With the cursor at the end of L2, press [,] [2nd] [alpha] [y] 1. Press [ENTER], and quadratic model has been defined in the equation editor as y1.

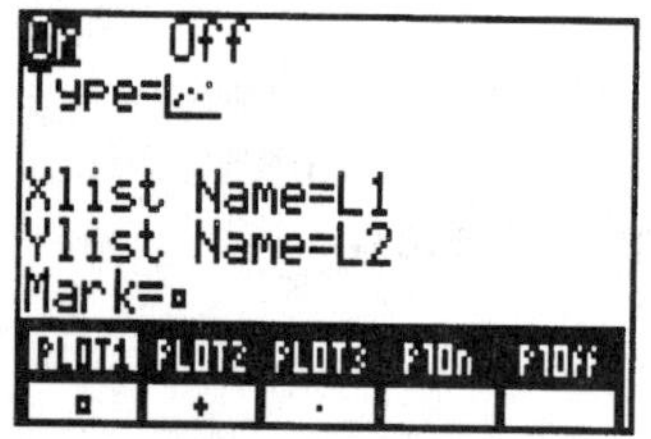

To see the graph of the regression equation together with the data, press [2nd] [STAT] [F3] [F1], press [ENTER] to turn plot1 On, and define the plot to be a scatter plot with lists L1 and L2 (see section 1.2 of this manual).

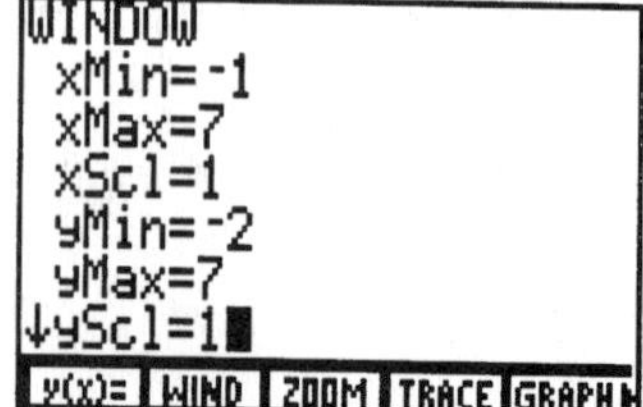

Set the graph viewing window ([GRAPH] [F2]) to $-1 \le x \le 7$ with xScl=1 by $-2 \le y \le 7$ with yScl=1.

Press $\boxed{F1}$ to access the equation editor, and if necessary press $\boxed{MORE}$ and then $\boxed{F3}$ (STYLE) as often as needed to select the graph style as Line.

To see the points and the graph of the model, press $\boxed{GRAPH}$ $\boxed{F5}$.

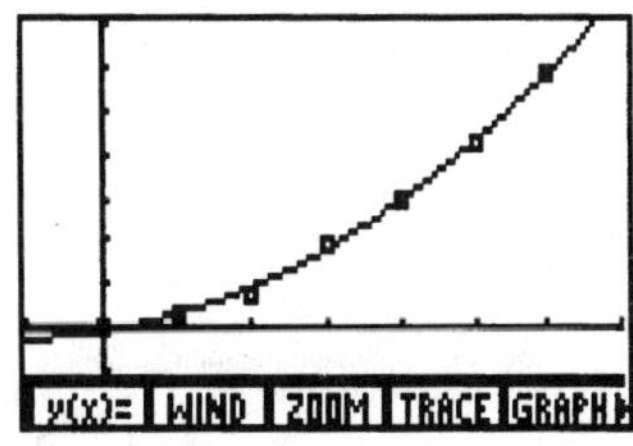

CHAPTER 2

SYSTEMS OF LINEAR EQUATIONS AND MATRICES

TI-86

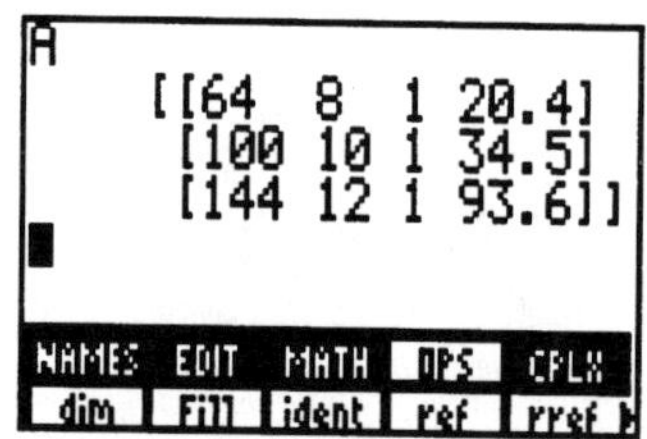

2.1 Systems of Two Linear Equations in Two Unknowns

In this section you will see how the TI-86 calculator can be used to solve systems of equations graphically. The calculator will be used to find the point of intersection of two graphs using TRACE and ZOOM. You will also be shown how to find a point of intersection of two graphs using the ISECT from the MATH menu in the graph editor. Using a table to find solutions to systems of equations will also be illustrated.

Solving Systems Graphically Using TRACE and ZOOM

Example 1 method 1 of this section in the text illustrated how to solve the system $\begin{aligned} x + y &= 3 \\ x - y &= 1 \end{aligned}$ graphically. Solve each equation for y to obtain $y = -x + 3$ and $y = x - 1$. Enter y1=-x+3 and y2=x-1 into the equation editor (GRAPH F1); be sure all Plots are turned off).

Set the viewing window to [−4, 4, 1] for x. You can allow the calculator to determine the y settings that best capture the graphs within this x range using ZFIT (zoom fit). Do this by pressing F3 (ZOOM) MORE F1 (ZFIT). The graphs of the equations are drawn with the y settings set by the calculator.

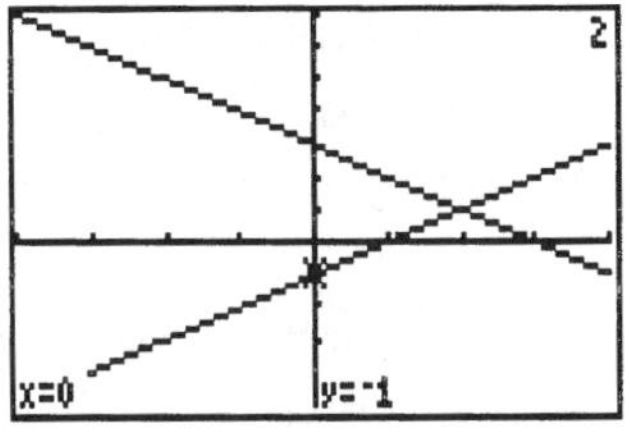

Press TRACE and ▼ several times and see the trace cursor "toggle" between the two graphs. You can know which graph you're tracing by the number in the upper right corner of the graph screen.

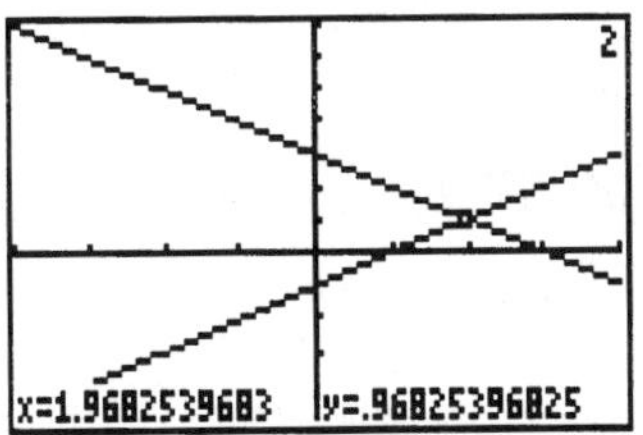

With the cursor on one of the graphs, use ▶ repeatedly to trace near the point of intersection. You can toggle again to see if there is visual movement of the cursor up and down. If not, then you have landed close to, or even on, the point of intersection (if the y values remain the same when you toggle).

To obtain a better approximation to the point of intersection, you can zoom in
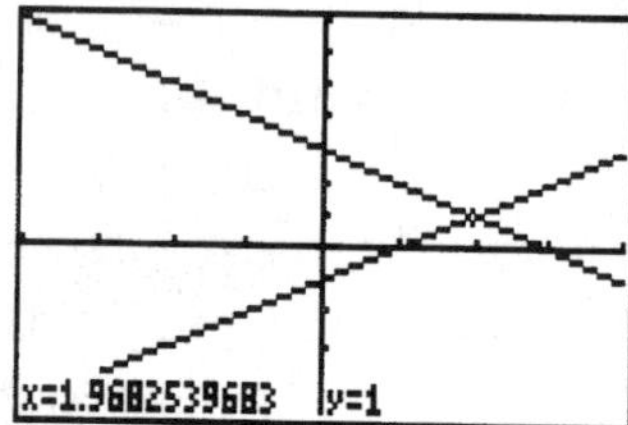
on the point using ZIN. With the cursor near the
point of intersection press GRAPH F3 (ZOOM) F2 (ZIN).
The original trace cursor (a box) will change to a free-
moving zoom cursor (a cross), giving you the freedom
to move the cursor anywhere on the screen for
marking the center of your zoom.

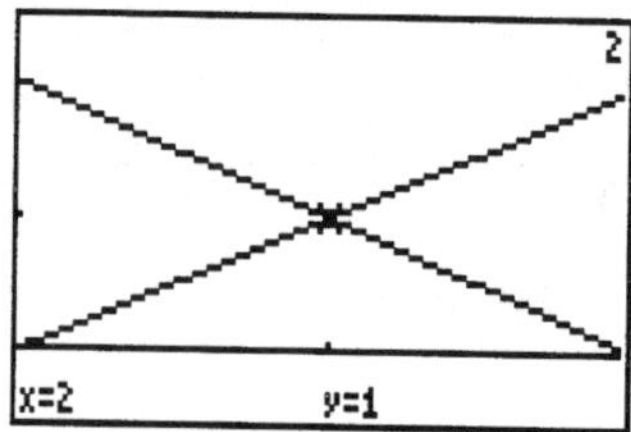
Since the cursor should be near the point of
intersection, press ENTER to view the "zoomed in"
graphs. When the zoom is complete, press GRAPH F4
and trace to the point of intersection. Toggling will
show that the point of intersection has been
approximated better (in fact, found exactly!).

Solving Systems Graphically Using ZOOM BOX

Zoom Box allows you to create a box around the point in which you would like
to zoom in on. Set the window back to [−4, 4] for x (GRAPH F2), and use ZFIT
(F3 MORE F1) to obtain the graphs in the original viewing window.

Press F3 (ZOOM) F1 (BOX) to obtain the zoom cursor.
Move the cursor to some spot upper left of the point
of intersection and press ENTER. Move the cursor
right and down to form a box around the point. Once
the box has been created, press ENTER and the graphs
will be redrawn in a new graph window defined by
the box you created.

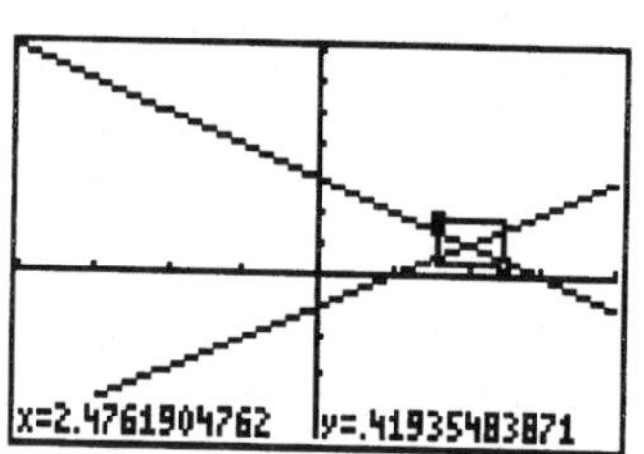

Press GRAPH F4 to trace near the point to obtain a
better approximation to the coordinates of the point
of intersection. Repeat this process as often as
needed to improve your approximation.

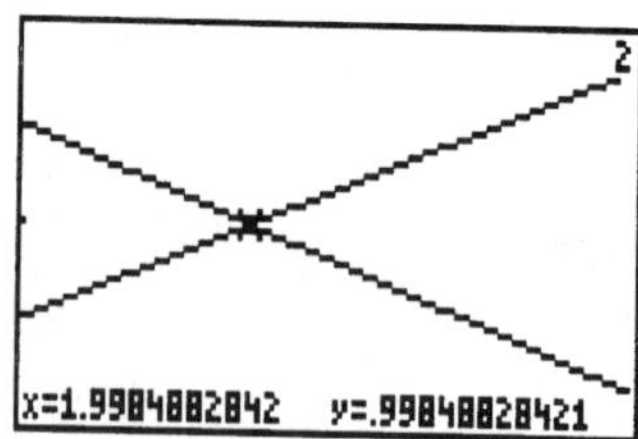

Solving Systems Graphically Using ISECT

Another method of finding the intersection of two graphs graphically is to use the ISECT feature of the GRAPH MATH menu. With the two equations defined in the equation editor, set the window to $-4 \leq x \leq 4$, and use ZFIT to obtain the graphs of the equations.

After the graphs have been drawn, press MORE F1 (MATH) to access the graph editors MATH menu. Press MORE to view the menu with the ISECT feature.

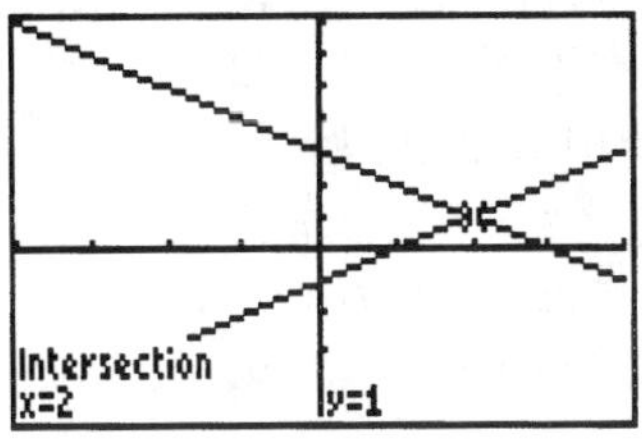

Press F3 (ISECT) to invoke this feature. Press ENTER to select y1 as the first curve; press ENTER again to select y2 as the second curve. At the guess prompt, arrow near the point of intersection, press ENTER and the calculator will display the coordinates of intersection.

Using a Table

First, graph the two equations and use TRACE to estimate that $x = 2$ might produce the point of intersection. Then press TABLE F2 to access the TABLE SETUP menu. Since $x = 2$ is an approximation, you can define TblStart=0 and ΔTbl=1. Set the Indpnt: option to Auto.

Press F1 (TABLE) to obtain a table of values for y1 and y2. You can view the table and observe that the y values of both equations is 1 when $x = 2$, indicating that the ordered pair $(2,1)$ is in fact the point of intersection and hence the solution to the system.

To use the TABLE feature in the manner used in the text, you would have to first set the Indpnt: option in TABLE SETUP to Ask. You would then access the table by pressing F1. At the prompt, enter the number 2 and press ENTER. Again you will see that both equations return a value of 1 for y and thus verifying that the point $(2, 1)$ is a solution.

Example 8 Finding a Break-Even Point Graphically on the TI-86
This is a great time to pull together some of the graphing calculator skills you have learned thus far. We will use this example from the text to do so. Here is an outline of the process for finding the break-even point, the point of intersection of the cost and revenue functions.

1. Identify the data points of the cost function as (3.5, 540) and (1, 650).
2. Enter these data points into lists L1 and L2 in the STAT editor, with the x coordinates in L1 and the corresponding y coordinates in L2. You may have to clear these lists of any existing data before entering the new values.
3. On the home screen, find the linear regression equation for the cost function, remembering to assign the regression equation to the equation editor (`LinR L1,L2,y1`).
4. Identify the data points of the revenue function as (2.99, 500) and (1.89, 600). Enter the x and y coordinates of these points in L3 and L4, respectively.
5. Repeat step 3 for the revenue function, assigning the regression equation to y2. Use `LinR L3,L4,y2`.
6. Set the window as [0, 4] on x by [0, 800, 100] on y.
7. Graph the cost and revenue functions, and find the point of intersection, indicating the break-even point.

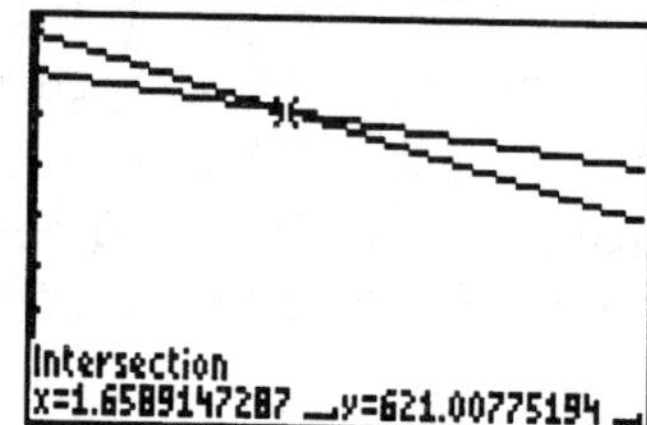

The point of intersection indicates that in order to break even, an access fee of $1.66 should be charged to obtain a needed revenue of $621.01.

2.2 Using Matrices to Solve Systems of Equations

In this section you will learn how to define (enter) a matrix into the TI-86.
You will then learn how to use TI-86 row operations on matrices for solving
systems of equations. A summary of matrices and row operations on the TI-
86 will be provided.

Defining a Matrix

We will use Example 7 *Row Operations Using Technology* from section 2.2 of
the text. To enter a matrix into the calculator, access the matrix editor via

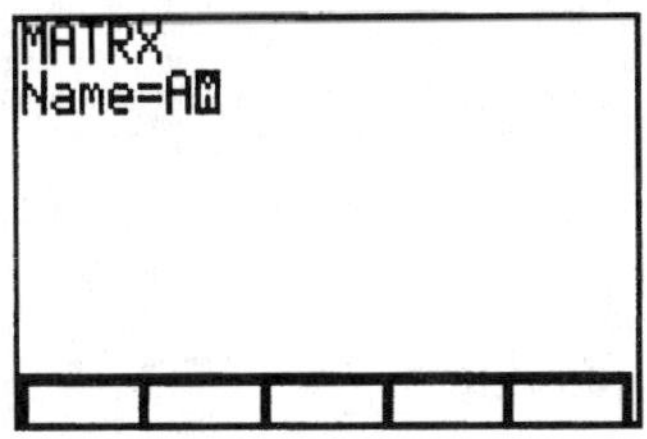

the MATRX menu by pressing [2nd] [MATRX] [F2] (EDIT).
The ALPHA-lock is on for you to assign a name to the
matrix you are about to define. Although matrix
names can be up to eight characters long, sticking to
single uppercase alpha names is usually most
convenient. Name the matrix A by pressing [A] [ENTER].

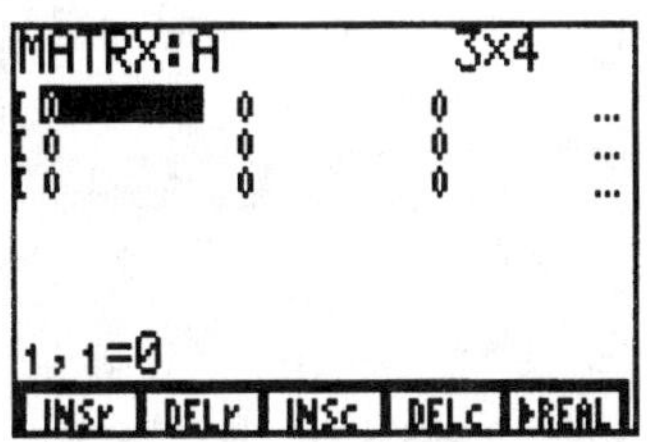

Press 3 [ENTER] 4 [ENTER] to define the size of A to be 3 by
4. You are now ready to enter the elements of A. At
the bottom of the screen you see the statement $1,1=0$.
This indicates that you are in row 1, column 1 and
that 0 is the current value of that particular entry.

Although the calculator will allow you to enter fractions such as 2/3 into the
matrix, it is often easier to clear the denominators before entering the values.
Do this on your paper before entering the any values in your matrix.

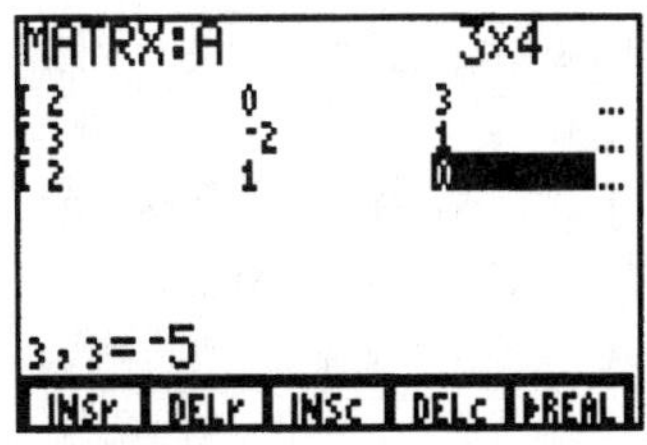

After clearing the denominators, press 2 [ENTER] to
enter the value 2 in row 1, column 1 and move to the
next entry. Continue to enter the remaining elements
of the matrix in the same manner. The screen to the
left shows some of the values
entered.

Once you are done entering the values, press [EXIT] to
return to the home screen. Then press [ALPHA] [A]
[ENTER] to view the matrix. The matrix can also be

found under and accessed through the NAMES in the MATRX menu.

Performing Row Operations on a Matrix

We are now ready to perform row operations on matrix A. The row operations are found in the matrix operations menu ([2nd] [MATRX] [F4] (OPS) [MORE]).

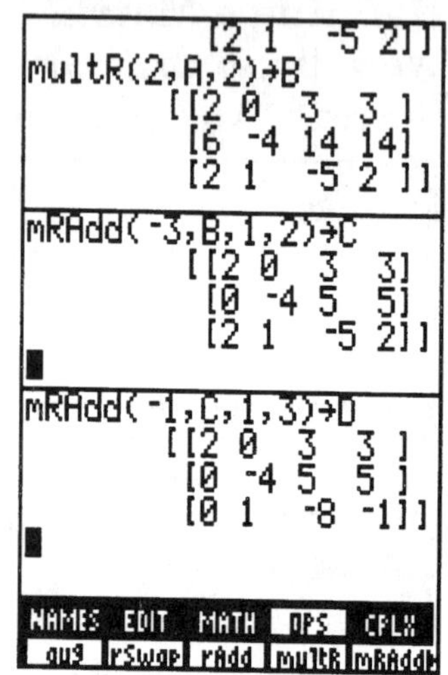

Using the entry value 2 of row 1, column 1 as the initial pivot, you want to replace R_2 with $2R_2 - 3R_1$. First multiply row 2 by 2, by pressing [F4] (multR) 2 [,] [ALPHA] [A] [,] 2 [)] [STO▶] [B] [ENTER] (the syntax for each row operation is given below). You should store each result as a different matrix so that if you make a mistake along the way, you will only have to repeat the previous matrix operation.

Now press [F5] (mRAdd) [(-)] 3 [,] [ALPHA] [B] [,] 1 [,] 2 [)] [STO▶] [C] [ENTER]. To complete the pivot, enter the expression mRAdd(-1,C,1,3)➔D and press [ENTER]. The entries below the pivot in column 1 will be zero.

Before we continue, let's take a brief look at the syntax for the calculators' row operations. A more detailed discussion of the operations will occur later in the summary of this section.

- rSwap(matrix name,row1,row2) interchanges the two rows specified
- rAdd(matrix name,row1,row2) adds two specified rows and places the result in row listed last
- multR(mulitple,matrix name,row) multiplies the row of the matrix by the specified mulitple
- mRAdd(multiple,matrix name,row1,row2) mulitplies the row listed first of the given matrix by the specified multiple, adds the result to second row listed, and then places the result in the row listed last

Be careful when entering your row numbers in these operations. For the operations rAdd and mRAdd, the calculator's syntax is basically (...,pivot row,change row), whereas the text goes by (...,change row,pivot row). For example, when the text says to perform $R_2 + R_3$, *row 2 is being changed* and

142

row 3 is the pivot row. On the calculator, however, you would enter
rAdd(A,3,2) so that the pivot row appears first and the change row last.

To complete reduction of the matrix in the example, you would do the
following.

1. $4R_3 + R_2$: multR(4,D,3)➔E [ENTER] and then rAdd(E,2,3)➔F [ENTER]
2. $9R_1 + R_3$: multR(9,F,1)➔G [ENTER] and then rAdd(G,3,1)➔H [ENTER]
3. $27R_2 + 5R_3$: multR(27,H,2)➔I [ENTER] and then mRAdd(5,I,3,2)➔J [ENTER]

After performing the third step, your screen should
look like the screen on the right. Thus, the result is
read as $(x, y, z) = \left(\dfrac{28}{18}, -\dfrac{140}{108}, -\dfrac{1}{27} \right)$, or $(x, y, z) =$

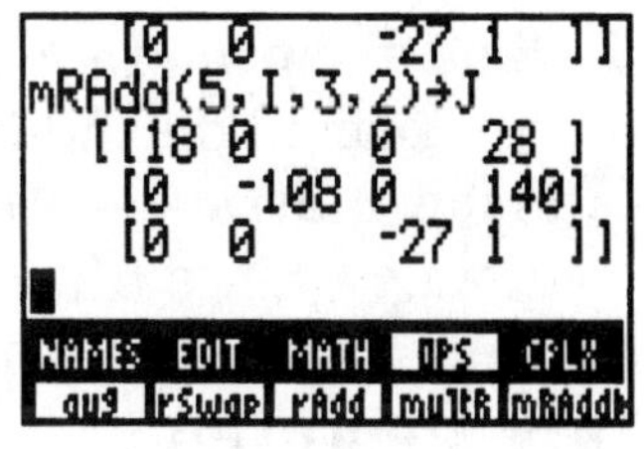

$\left(\dfrac{14}{9}, -\dfrac{35}{27}, -\dfrac{1}{27} \right)$ in reduced form.

One last comment regarding this example. You can use the TI-86 to take the
above process four steps further to view the solutions directly on your
calculator's screen! Doing so only requires dividing each row by its pivot.
Thus, do the following.

- $\dfrac{1}{18}R_1$: multR(1/18,J,1)➔K [ENTER]

- $-\dfrac{1}{108}R_2$: multR(-1/108,K,2)➔L [ENTER]

- $-\dfrac{1}{27}R_3$: multR(-1/27,L,1)➔M [ENTER]

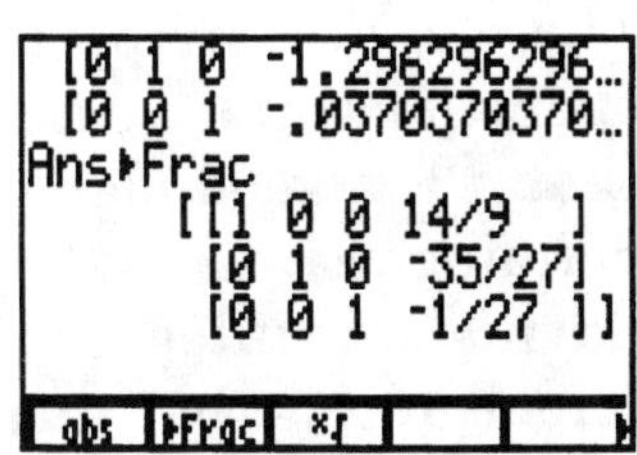

- Press [CUSTOM] [F2] (▶Frac) to convert the entries of
  the final matrix to fractions, and there you have
  it! The solutions are right there in the right last
  column!

SUMMARY OF MATRICES FOR SECTION 2.2 ON THE TI-86

About Matrices on the TI-86

1. Matrices should be entered in the `MATRX` editor (`2nd` `[MATRX]` `F2`). Matrices can be referred to in `NAMES` under the `MATRX` menu. However, if you use single uppercase letters to name matrices, they can easily be referred to on the home screen by that letter.
2. When in the `MATRX` editor, press `EXIT` to return to the home screen.
3. Perform matrix operations, including row operations, using the operations found in `MATRX OPS`.
4. The algebra of matrices (adding, subtracting, multiplying, etc.) is performed on the home screen.
5. View matrices with decimal or fractional entries on the home screen by using `▶Frac` after the matrix name.

Row Operations

In the Gauss-Jordan matrix reduction method, there are four row operations used. They are `rSwap`, `rAdd`, `multR`, and `mRAdd`. When performed on a matrix, these row operations do not change the original matrix. Rather, a new matrix is developed from the matrix the operation is being performed on.

Thus, when using `STO▶` to store a matrix resulting from a row operation, it is highly recommended that the matrix be stored using a name different from the original matrix. This preserves the original matrix and keeps you from having to re-enter the matrix if needed later.

There are two important things to remember when performing row operations on matrices using the TI-86 row operation commands. The first is that it is a good idea to always clear denominators in the matrix before defining the matrix in the calculator. This allows you to begin with a matrix having only integer entries rather than fractional. The second, and most important, is that *except for the operation* `rSwap`, *the last row listed in the row operation command is the row that will be changed.*

The row operations are found in the matrix operations menu `MATRX OPS`. Each of the four row operations is described below with their syntax, purpose, and an example. Following the example is the procedure for using the operation to perform the command given in the example.

rSwap(
- **Syntax:** rSwap(matrix name,row i,row j)
- **Purpose: Interchange (swaps) row i with row j**
- **Example: rSwap(A,2,3)→B swaps row 2 and row 3 of matrix A and stores the result as matrix B**

Assuming matrix A has been defined, we carry out the operation in the example above as follows.
1. On the home screen, access the matrix operations menu MATRX OPS by pressing 2nd [MATRX] F4 (OPS) MORE. The menu should appear and remain on the screen until exited or a calculator error (command entry, syntax, etc.) is performed.
2. Press F2 to select the rSwap operation.
3. Enter ALPHA [A] , 2 , 3) STO► [B] ENTER to see the result.

rAdd(
- **Syntax:** rAdd(matrix name,row i,row j)
- **Purpose: Adds row i to row j and places the result in row j**
- **Example: rAdd(A,1,3)→B takes row 1 and row 3 of matrix A, adds them together, places the answer in row 3, and then stores the new matrix as matrix B**

Assuming matrix A has been defined, we carry out the operation in the example above as follows.
1. If the MATRX OPS menu is not currently shown on the home screen, access it by referring to step 1 of the rSwap procedure given above.
2. Press F3 to select the rAdd operation.
3. Enter ALPHA [A] , 1 , 3) STO► [B] ENTER to see the result.

multR(
- **Syntax:** multR(constant multiplier,matrix name,row i)
- **Purpose: Multiplies row i by a constant and then replaces row i with the result**
- **Example: multR(-4,A,3)→B multiplies row 3 of matrix A by –4, replaces row 3 with the answer, and then stores the new matrix as matrix B**

Assuming matrix A has been defined, we carry out the operation in the example above as follows.
1. If the MATRX OPS menu is not currently shown on the home screen, access it by referring to step 1 of the rSwap procedure given above.
2. Press F4 to select the multR operation.

3. Enter [(-)] 4 [,] [ALPHA] [A] [,] 3 [)] [STO▸] [B] [ENTER] to see the result.
4. If the resulting matrix has decimal entries (this can happen if your constant multiplier is a decimal or a fraction), you can press [CUSTOM] [F2] (▸Frac) [ENTER] to view the new matrix with its fractional entries. Doing this, however, will replace the MATRX OPS menu with the CUSTOM menu.

mRAdd(

- Syntax: mRAdd(constant multiplier,matrix name,row i,row j)
- **Purpose: Multiplies row i by a constant, adds the result to row j, and places row j with the answer**
- **Example: mRAdd(2/3,A,2,1) multiplies row 2 of matrix A by 2/3, adds the result to row 1, replaces row 1 with the answer to this addition, and then stores the new matrix as matrix B**

Assuming matrix A has been defined, we carry out the operation in the example above as follows.
1. If the MATRX OPS menu is not currently shown on the home screen, access it by referring to step 1 of the rSwap procedure given above.
2. Press [F5] to select the mRAdd operation.
3. Enter 2 [÷] 3 [,] [ALPHA] [A] [,] 2 [,] 1 [)] [STO▸] [B] [ENTER] to see the result.
4. If the resulting matrix has decimal entries (this can happen if your constant multiplier is a decimal or a fraction), you can press [CUSTOM] [F2] (▸Frac) [ENTER] to view the new matrix with its fractional entries. Doing this, however, will replace the MATRX OPS menu with the CUSTOM menu.

REF AND RREF USING THE TI-86

The TI-86 can reduce a matrix to *row echelon form* using the command **ref** and to *reduced row echelon form* using the command **rref**. Like the matrix row operation commands, the original matrix is not changed when these commands are used. Therefore the resulting matrix should be stored using a different name for future reference while at the same time preserving the original matrix.

Row Echelon Form (ref)

- **Syntax: ref *matrix name***
- **Purpose: Computes a row echelon form matrix for a given matrix**
- **Example: ref A→B returns a row echelon form matrix for matrix A and stores the resulting ref matrix as matrix B**

Assuming matrix A has been defined, we carry out the operation in the example as follows.

1. Access the MATRX OPS menu ([2nd] [MATRX] [F4] (OPS)).
2. Press [F4] (ref) [ALPHA] [A] [ENTER] to see the result.
3. If the resulting matrix has decimal entries, you can press [CUSTOM] [F2] (►Frac) [ENTER] to view the new matrix with its fractional entries.

Reduced Row Echelon Form (rref)

- **Syntax:** rref *matrix name*
- **Purpose: Computes the reduced row echelon form matrix for a given matrix**
- **Example:** rref A➔B **returns** *the* **reduced row form matrix for matrix A and stores the resulting rref matrix as matrix B**

Assuming matrix A has been defined, we carry out the operation in the example as follows.

1. Access the MATRX OPS menu ([2nd] [MATRX] [F4] (OPS)).
2. Press [F5] (rref) [ALPHA] [A] [ENTER] to see the result.
3. If the resulting matrix has decimal entries, you can press [CUSTOM] [F2] (►Frac) [ENTER] to view the new matrix with its fractional entries.

See "You're the Expert" later in this chapter for an example of how rref can be used to solve problems.

2.3 Applications of Systems of Linear Equations

There are no new calculator skills presented in this section.

You're the Expert – The Impact of Regulating Sulfur Emissions

Here you are given the task of giving projections of the annual cost to utilities and the cost in jobs if the EPA regulations requiring a 15 million ton rollback in sulfur emissions were to be enacted. Systems of equations were used to find quadratic models for the cost to utilities, C, and the cost in jobs, J, in terms of the rollback tonnage t. Here you will learn how to use matrices and the $\texttt{rref}$ command to solve the systems of equations that were developed from the data for determining the coefficients of the quadratic models.

Solving a system of equations

The system representing the quadratic model for the cost to utilities was found to be

$$20.4 = 64a + 8b + c$$
$$34.5 = 100a + 10b + c .$$
$$93.6 = 144a + 12b + c$$

Enter the 3 by 4 augmented coefficient matrix $A = \begin{bmatrix} 64 & 8 & 1 & 20.4 \\ 100 & 10 & 1 & 34.5 \\ 144 & 12 & 1 & 93.6 \end{bmatrix}$ into

your calculator as matrix A. Exit to the home screen and view matrix A to verify that the matrix has been inputted correctly.

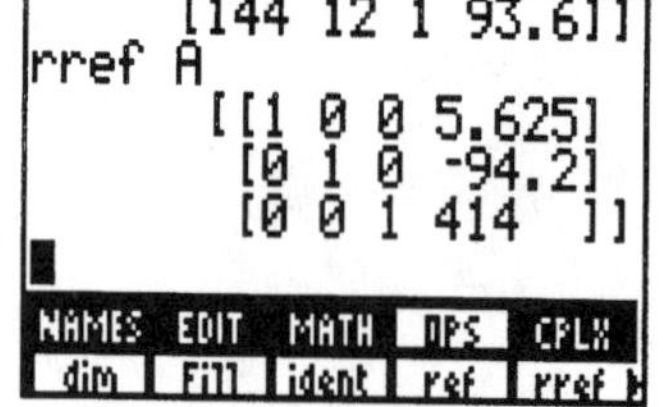

Press [2nd] [MATRX] [F4] (OPS) [F5] (rref) [ALPHA] [A] [ENTER] and the solution to the system is immediately determined to be $(a,b,c) = (5.625, -94.2, 414)$. Thus, the cost equation for utilities becomes $C = 5.625t^2 - 94.2t + 414$. A similar method can be used to determine the equation for J, the cost in jobs.

CHAPTER 3

MATRIX ALGEBRA
AND
APPLICATIONS

TI-86

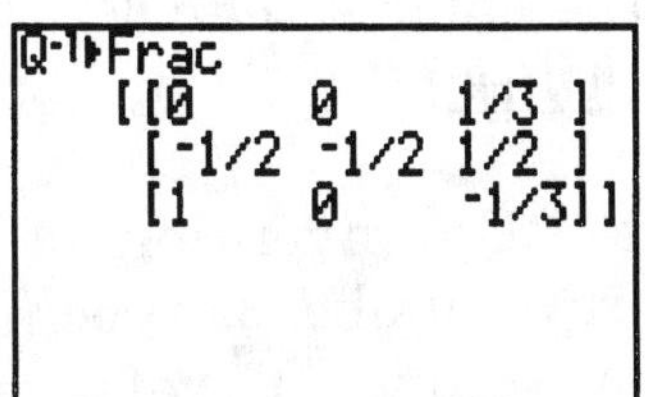

3.1 Matrix Addition and Scalar Multiplication

In this section you will learn how you can refer to and edit the entries of a matrix on the TI-86. You will also learn how to add and subtract scalar multiples of matrices. How to obtain the transpose of a matrix will also be illustrated.

An introduction to matrices on the TI-86 was presented in section 2.2 of the previous chapter. There you learned how to define a matrix into the calculator, as well as a method of how matrices can be used to solve linear systems of equations. You also learned how to navigate around a portion of the TI-86's matrix editor, in particular the matrix operations menu MATRX OPS. This chapter continues the discussion on matrices, including the algebra of matrices.

Referring to and Editing the Entries of a Matrix

Consider 2×3 matrix $A = \begin{bmatrix} -1 & 9 & 17 \\ 4 & 6 & -13 \end{bmatrix}$. The entry in the first row, second column is 9, and we denote this by $a_{12} = 9$. Similarly, $a_{11} = -1$, $a_{21} = 4$, and $a_{23} = -13$. Can you identify the values of the remaining entries a_{13} and a_{22}?

Enter this matrix into your calculator, naming it A (see section 2.2 – *Defining a matrix* and/or *About matrices on the TI-86*). After you have defined A in your calculator, return to the home screen (EXIT) and view the matrix (ALPHA [A] ENTER).

The TI-86 follows the same "RC" convention as mentioned in the text: the row number is specified first and the column number second. To refer to the 1,2-entry of A (a_{12}) on your calculator, press ALPHA [A] (1 , 2) ENTER. Then press 2nd ENTER to recall the last command issued, change the 1 to 2, the 2 to 3 and press ENTER to find a_{23}.

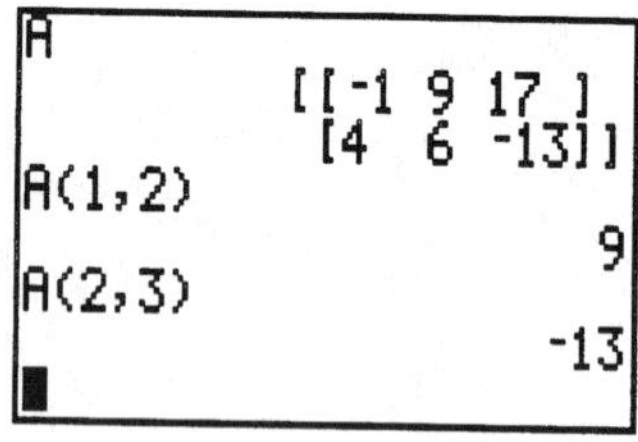

Continue in this manner and see if the values you expected for a_{13} and a_{22} were correct.

Suppose after viewing matrix A on the home screen, you realize that the 1,3-entry was not 17, but 47. Then you would need to edit the matrix. One way of doing so is obvious — go back into the matrix editor and overwrite the entry 17 with 47. But there is another option.

Press $\boxed{\text{CLEAR}}$ to clear your home screen. This is not necessary, but it is often good to start a series of commands with a clear home screen. Recall matrix A on the home screen ($\boxed{\text{ALPHA}}$ [A] $\boxed{\text{ENTER}}$). Now press 47 $\boxed{\text{STO▸}}$ [A] $\boxed{\text{ALPHA}}$ $\boxed{(}$ 1 $\boxed{,}$ 3 $\boxed{)}$ $\boxed{\text{ENTER}}$. Now view matrix A again and verify that the change of this entry from 17 to 47 has been successfully made!

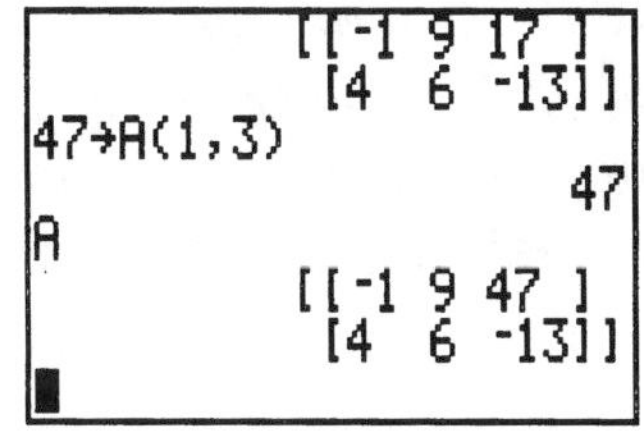

You can perform this procedure to edit any entry of a matrix from the home screen. Try changing the a_{22} entry from 6 to 109.

Adding and Subtracting Scalar Multiplies of Matrices

Any matrix can be multiplied by a scalar multiple c on the TI-86. However, not any two matrices can be added and subtracted. Only matrices that have the same dimension (size) can be added or subtracted.

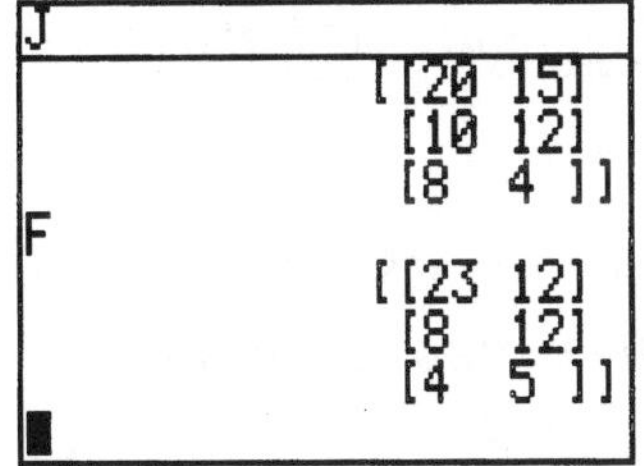

Let's consider Examples 2 and 3 (*Sales*) from section 3.1 of the text. Enter the matrices representing the January and February sales for the APlus auto parts store into your calculator as J and F, respectively. Be sure to view each matrix on the home screen to make certain that each one has been defined correctly.

At the home screen, press $\boxed{\text{ALPHA}}$ [F] $\boxed{-}$ $\boxed{\text{ALPHA}}$ [J] $\boxed{\text{ENTER}}$ to obtain the same result as given in the text.

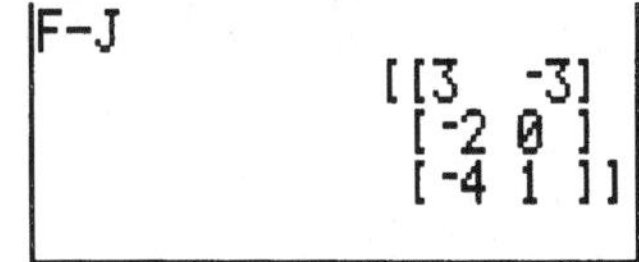

Example 3 *Sales* presented in the text gives the January sales of the store, in Canadian dollars, as

$$A = \begin{bmatrix} 140{,}000 & 105{,}000 \\ 30{,}000 & 36{,}000 \\ 96{,}000 & 48{,}000 \end{bmatrix}.$$

Let's veer just a bit from the example as given in the text. Currently, the Canadian dollar is worth about $0.69US, so to compute the revenue in current US dollars we would have $0.69A$. See the calculator screen for the result.

To evaluate the expression $2F + 3J$, enter the expression on the home screen in the same order that you see it written here.

Transposing a Matrix

To transpose matrix J on the TI-86, we need the transpose symbol "^{T}". This is found in the MATRX MATH menu.

Clear your home screen. Press [2nd] [MATRX] [F3] (MATH) to access this menu. Now press [ALPHA] [J] [F2] (^{T}) [ENTER] to view the transpose of J.

Be aware that for editing purposes, we will sometimes represent column matrices as the transpose of a row matrix. For example, we may write $[5 \ -4 \ 2]^T$ for the column matrix $\begin{bmatrix} 5 \\ -4 \\ 2 \end{bmatrix}$.

3.2 Matrix Multiplication

In this section you will be shown how to use your TI-86 to perform matrix
multiplication, including some details that you must pay attention to. You
will also see how to define an identity matrix of any size. An exercise from
the text will be used to illustrate the use of these calculator skills.

Matrix Multiplication – Some Details

Matrix multiplication is a simple task for the TI-86. In the previous section,
you learned how easy it was to perform scalar multiplication with the
calculator. For example, if you want to multiply some matrix, say A, by 15,
you simply define A in the matrix editor, exit to the home screen, enter 15A
and press ENTER to view the result. However, when multiplying matrices on
the TI-86, there are a couple of issues you need to be careful of and watch out
for.

The first is a technical issue. Whenever multiplying matrices with the TI-86,
you must use the ∗ symbol (×). Suppose, for example, that you want to
multiply matrices $A = \begin{bmatrix} 2 & 7 \\ 3 & -5 \end{bmatrix}$ and $B = \begin{bmatrix} 1 & 1 \\ 4 & -2 \end{bmatrix}$.

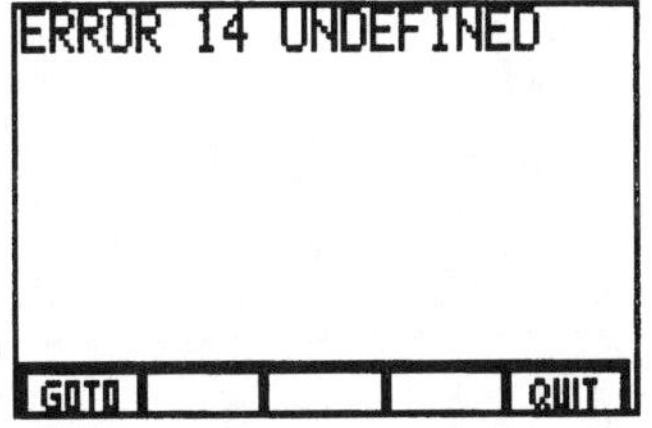

Define these two matrices in your calculator. On your
home screen, enter the expression AB (ALPHA [A]
ALPHA [B]) for the product of A and B, and press ENTER.
You will receive something that you may not have
expected, most probably the error message shown in
the calculator screen on the left.

Note that you are given two options on this screen:
GOTO (F1) and QUIT (F5). Press F1 and the
calculator will return you to the home screen, with
the prompt blinking near the location of the error.
Arrow to the left and place the cursor over the B,
press 2nd [INS] × to insert the ∗ symbol, and press
ENTER to view the result of A∗B.

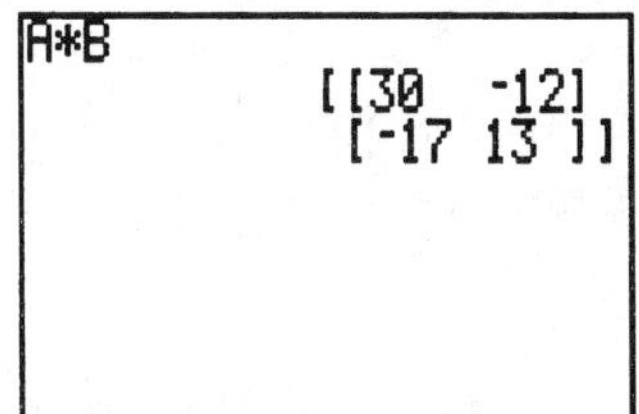

When you enter AB on the home screen and press $\boxed{\text{ENTER}}$, the calculator doesn't know that you are trying to multiply two matrices. The TI-86 thinks that you are asking to view the contents (a value, matrix, etc.) of the memory register named "AB". If there is something in the register, it will return its content, but it won't be the product you are looking for. Otherwise, it will return the error message shown above, indicating that AB has not been defined.

The second issue is theoretical. When comparing the dimensions of the matrices you want to multiply, the number of *columns* of the first matrix (the matrix on the left) *must* be the same and the number of *rows* of the second matrix (the matrix on the right).

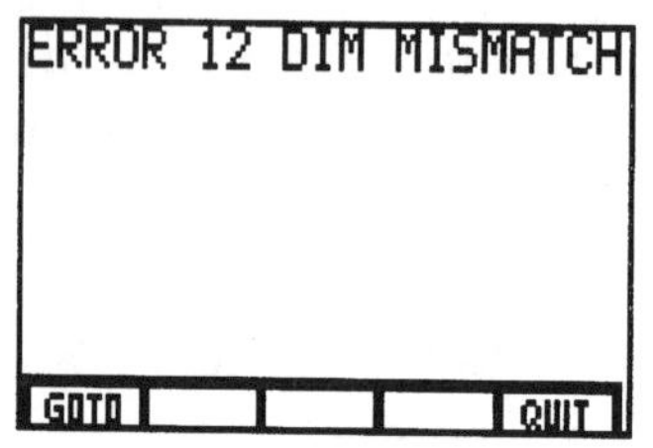

If you attempt to multiply two matrices whose dimensions violate this principle, you will receive the error message shown in the accompanying graphing calculator screen. To fix this, choose QUIT ($\boxed{\text{F5}}$) and make the necessary corrections to ensure that the dimension rule for matrix multiplication is satisfied.

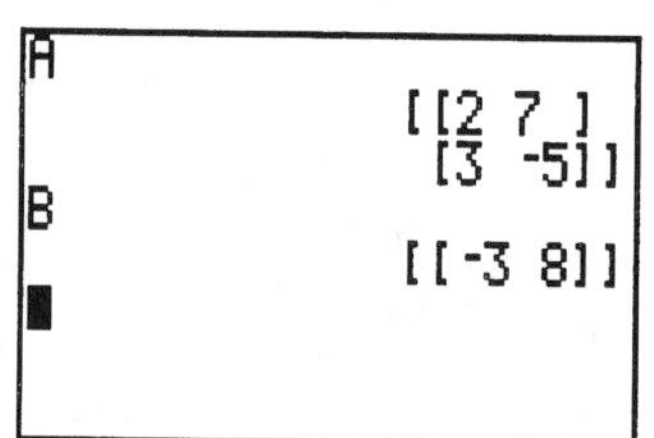

For example, change matrix B above to be the 1x2 matrix $B = [-3, 8]$. Enter the product $A*B$ on the home screen and press $\boxed{\text{ENTER}}$. The error message shown above will appear.

Press $\boxed{\text{F5}}$ (QUIT), and when the calculator returns to the home screen enter the product B*A and press $\boxed{\text{ENTER}}$. Notice that now the order of the matrices is such that the row and column size requirement stated above is satisfied, and so a result was returned.

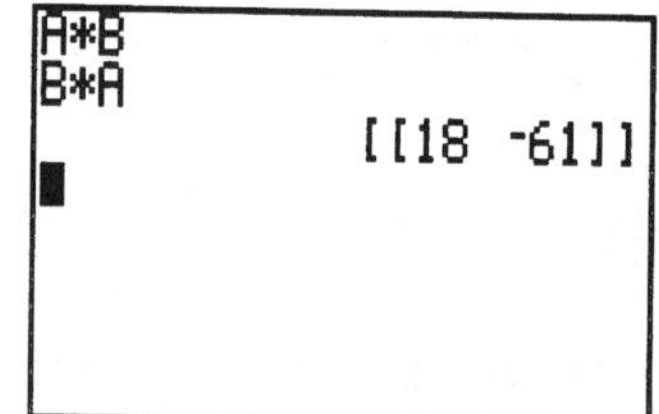

The Identity Matrix

The identity matrix I can be defined in the matrix editor just like any other matrix. But there is an alternative way of defining an identity matrix on the

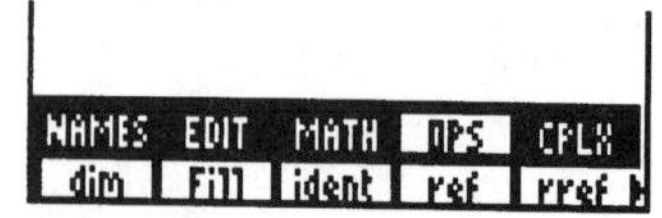

TI-86. In the MATRX OPS menu, there is an operation called ident.

The syntax for this operation is ident n, where n represents the size of the identity matrix. For example, ident 2 will return a 2×2 identity matrix, and ident 4 will return a 4×4 identity matrix.

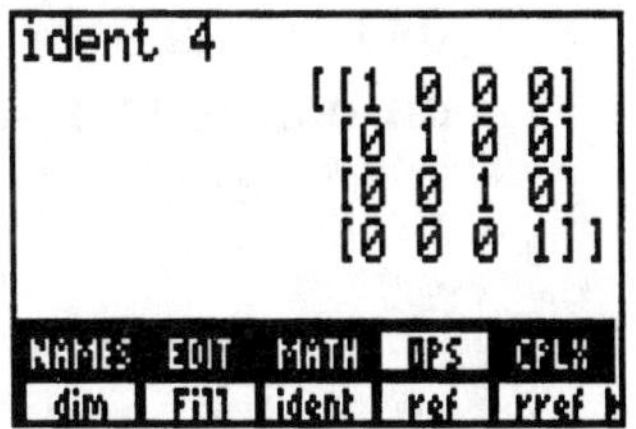

You can use this operation directly in calculations on the home screen. For example, if A is a 2×2 matrix defined in your calculator, then computing ident 2*A and A*ident 2 on the home screen will illustrate the property $IA = AI$ for any square matrix.

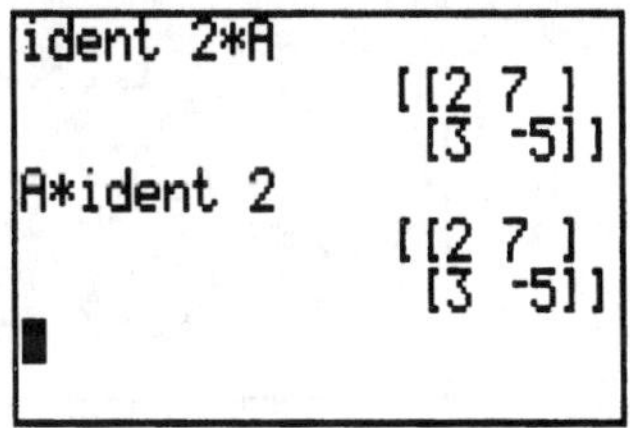

You can also define the identity matrix on the home screen as I by pressing 2nd [MATRX] F3 (ident) 2 (or any other number) STO▶ [I] ENTER. Doing this allows you to verify $IA = AI$ by using I*A and A*I.

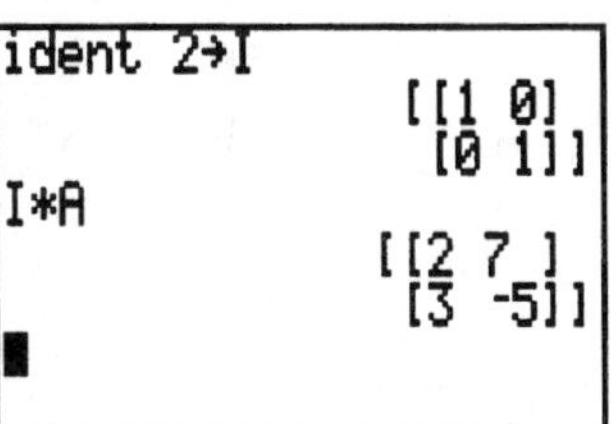

Applications – Exercises From the Text

55. Revenue is determined by the price per unit times the quantity sold. Define the price matrix P as a row matrix (or column matrix) and the quantity matrix Q as a column matrix (or a row matrix if you chose the units matrix to be a column matrix) in your calculator. On the home screen, compute $P*Q$. Thus the total revenue for the week was $1,510.

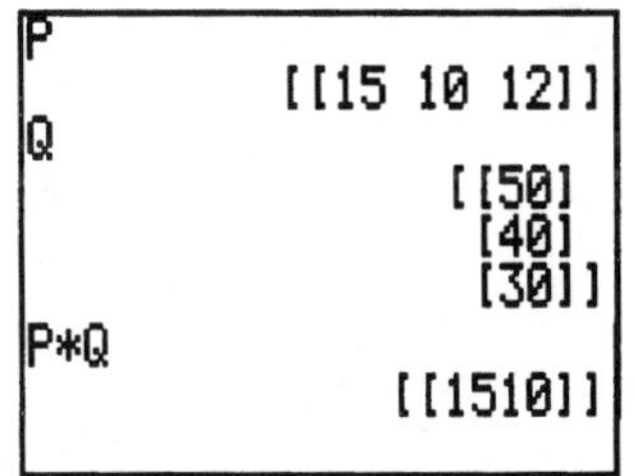

69. Input P and the 1987 distribution as a row vector A into your calculator. View the matrices on the home screen to verify correct entry. See the screen below, left.

    Press 2nd [MODE] ▼ ▶ ▶ ENTER to change the decimal mode setting to 1. Press EXIT to return to the home screen. See the screen below, center.

Now enter $A*P$ and press ENTER to see the result, using the arrow keys to scroll right and left as necessary. You can conclude from the result that in 1988, the population breakdown was 49 million in the Northeast, 58.9 million in the Midwest, 58.7 million in the South, and 48.2 million in the West. See the screen below, right.

Be sure to return the decimal mode setting back to Float before moving on in this manual.

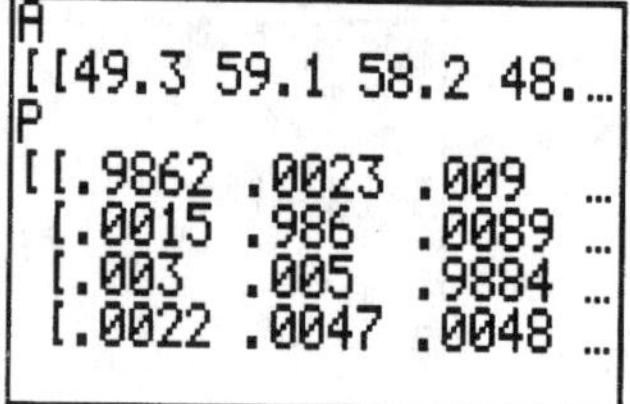 

3.3 Matrix Inversion

Here you will learn how to use the TI-86 to find the inverse of a square matrix, how to use find the determinant of a square matrix, and how to solve systems of n linear equations in n unknowns using the inverse.

Finding the Inverse of a (Square) Matrix

If you look at the key directly above the number $\boxed{7}$ on your keypad, you will see the primary function key $\boxed{\text{EE}}$ and its secondary function $[x^{\text{-}1}]$. This secondary key is needed to find the inverse of a square matrix.

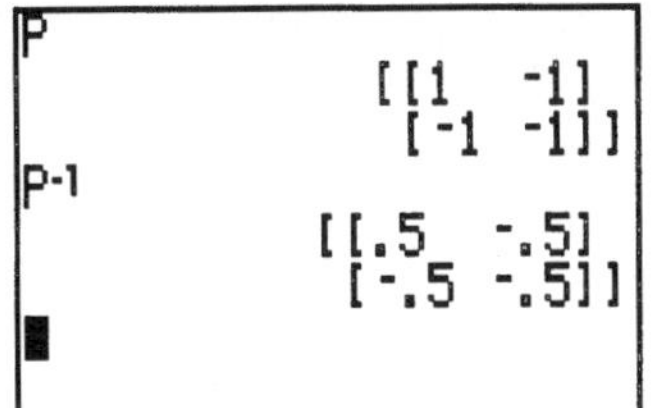

Using part (a) of Example 2 *Computing the Inverse* in section 3.3 of the text, let's find P^{-1} using the TI-86. Input matrix P into your calculator. After viewing P on the home screen, press $\boxed{\text{ALPHA}}$ [P] $\boxed{\text{2nd}}$ $[x^{\text{-}1}]$ $\boxed{\text{ENTER}}$. The inverse of P will appear on your screen.

Noting that there are decimal entries in the inverse, press $\boxed{\text{CUSTOM}}$ $\boxed{\text{F2}}$ ($\blacktriangleright$Frac) $\boxed{\text{ENTER}}$ to obtain fractional entries for A^{-1}. You were shown how to set your custom menu in section 1.3 of this manual.

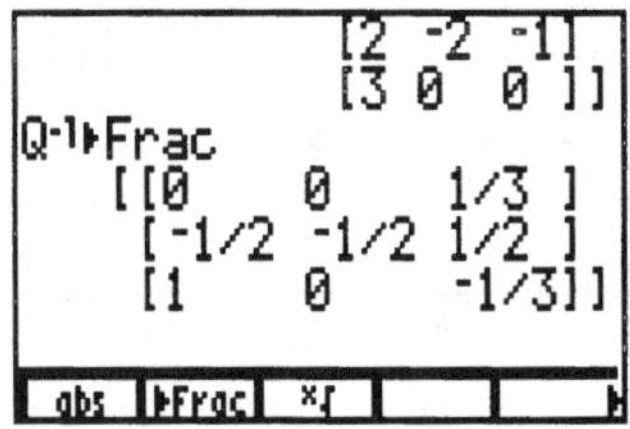

Most times inverse matrices will have fractional entries. Knowing this, we can enter $A^{-1}\blacktriangleright$Frac all at once to obtain fractional entries for A, if any. Try this with part (b) of Example 2. Remember to view matrix Q for errors before finding its inverse.

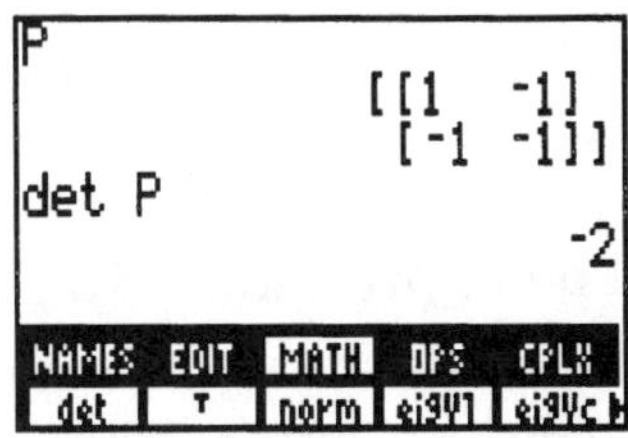

Finding the Determinant of a Matrix

The determinant function is found in the MATRX MATH menu. To compute the determinant of matrix P of Example 2 above, press $\boxed{\text{2nd}}$ [MATRX] $\boxed{\text{F3}}$ (MATH) $\boxed{\text{F1}}$ (det) $\boxed{\text{ALPHA}}$ [P] $\boxed{\text{ENTER}}$. Note that a non-zero value was returned. This indicates that P is *invertible* (i.e., non-singular).

Consider matrix A of Example 1 *Singular Matrix* in the text. Input this matrix into your calculator, naming it A. View A, and then find its determinant. Since the determinant is zero, matrix A is a *singular* matrix (and hence not invertible). You can see this by the error message you obtain when attempting to use the TI-86 to compute the inverse A.

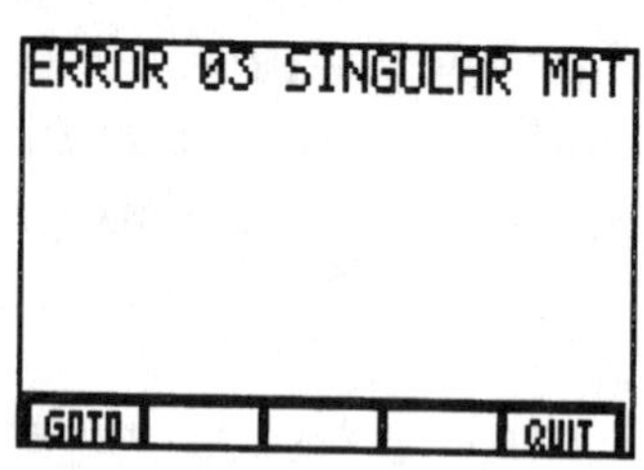

Using the inverse to solve n linear systems of n equations
Part (a) of Example 4 *Solving Multiple Systems Using an Inverse* asks you to solve the system

$$2x \quad\ + z = 1$$
$$2x + y - z = 1$$
$$3x + y - z = 1$$

using an inverse.

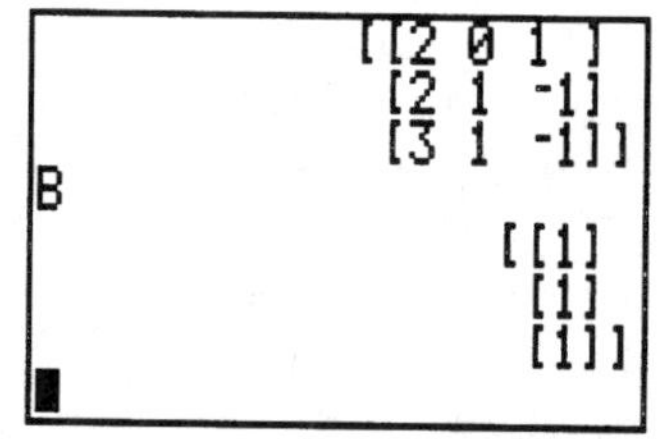

Input the coefficient matrix $A = \begin{bmatrix} 2 & 0 & 1 \\ 2 & 1 & -1 \\ 3 & 1 & -1 \end{bmatrix}$ and the constant matrix $B = \begin{bmatrix} 1 \\ 1 \\ 1 \end{bmatrix}$ into your calculator. View these matrices to verify that they have been defined correctly.

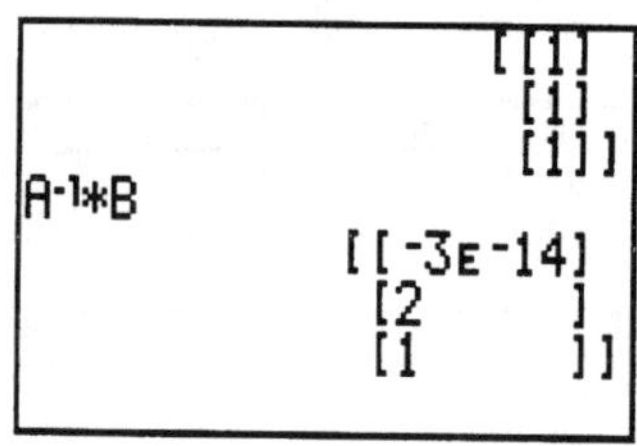

At the home screen, enter $A^{-1}*B$ and press [ENTER] to see the result. The answer matrix looks a bit different from the one given in the text. The first entry shows the value $-3E-14$ and not 0.

The expression $-3E-14$ is the calculators way of representing a number in scientific notation. This expression is therefore -3×10^{-14}, which in decimal form is -0.00000000000003. This number is fairly close to zero as far as we are concerned. In general, an expression such as this with a double-digit *negative* exponent, we treat this number as 0. Hence, our solution matrix is $[0 \quad 2 \quad 1]^T$.

Application – Exercises from the Text

59. Since the encrypted matrix has 14 columns, you will need to re-define it to be a matrix E to have dimension 2×7. Thus

$$E = \begin{bmatrix} 33 & 54 & 11 & 20 & 29 & 65 & 41 \\ 69 & 126 & 27 & 60 & 59 & 149 & 87 \end{bmatrix}.$$

Input matrices A and E into your calculator and then view them for correctness. When inputting E, be sure to recognize that the first row is made up of the entries in odd positions, and the second row is the entries in the even positions.

Therefore, when inputting 33 in the first row, first column and pressing ENTER, the e_{12} entry should be 54, the value in the 3rd position of the encrypted row matrix, and so on.

At the home screen enter $A^{-1}*E$ to obtain the original message. The original message is given by the row matrix
$[3 \quad 15 \quad 18 \quad 18 \quad 5 \quad 3 \quad 20 \quad 0 \quad 1 \quad 14 \quad 19 \quad 23 \quad 5 \quad 18]$. Hence the message is CORRECT ANSWER.

3.4   Input-Output Models

There are no new calculator skills for this section. However, let's outline how to solve Example 3 *Kenya Economy* from the text.

1. Use the `MATRX` editor to enter the technology matrix A into your calculator. We suggest that you write the matrix on paper first with the appropriate rounding of the entry quotients *column entry/column total* and then enter the values into the calculator.

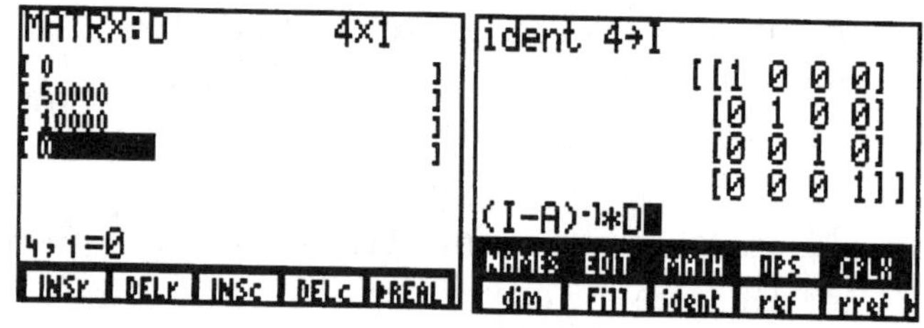

2. Enter matrix D^+ as matrix D.

3. On the home screen compute $(I-A)^{-1}*D$. Round the results to the nearest integer (why?).

4. You can press [2nd] [ENTER] to recall the last entry and edit the command line to compute $(I-A)^{-1}$ separately if needed. Before doing so, set the decimal mode to 5 decimal places ([2nd] [MODE]).

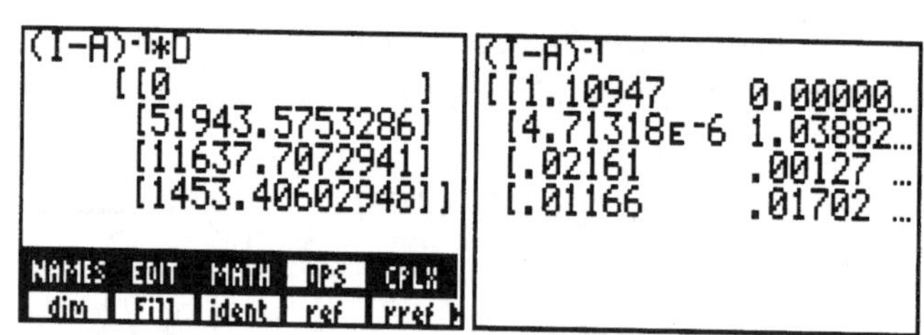

Remember to return the decimal mode back to `Float` when finished.

Your're the Expert – The Japanese Economy

There are no new calculator skills in this section.

CHAPTER 4

LINEAR PROGRAMMING

TI-86

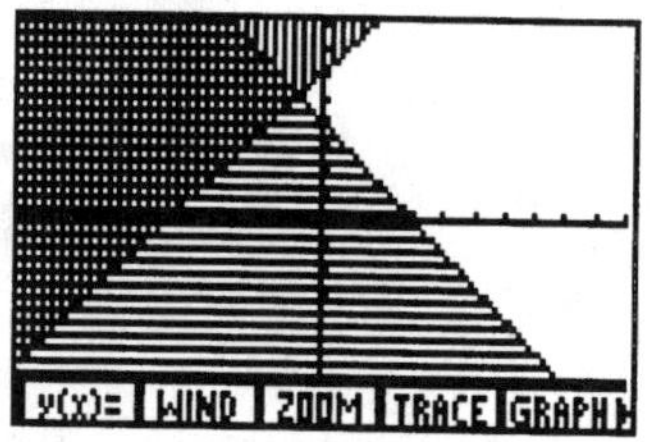

4.1 Graphing Linear Inequalities

In this section you will learn how to use the TI-86 to sketch the solution region represented by a linear inequality in two variables. You will also learn how to sketch the feasible region of a system of (simultaneous) linear inequalities.

To use the TI-86 to sketch the graph of a linear inequality, you must first solve for y and then determine which side of the inequality (or, which half-plane) represents the solution set. Remember, as mentioned in the text, *the part you do not want will be shaded, leaving the solution region blank.*

Let's see how this is done for the inequality $2x + 3y \leq 6$. Solve the equation $2x + 3y = 6$ for y to obtain $y = -\dfrac{2}{3}x + 2$. Access the equation editor (GRAPH F1) and press F4 to delete any equations currently in the editor. Define this equation as y1.

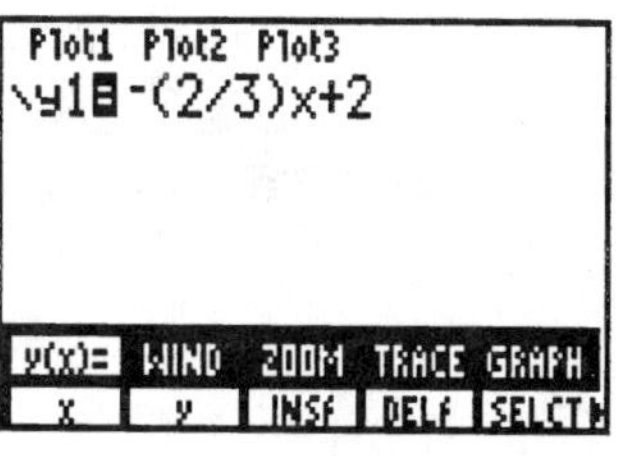

Set the graph viewing window (GRAPH F2) to the standard settings: $[-10, 10]$ on x by $[-10, 10]$ on y. Then press F5 to graph the equation.

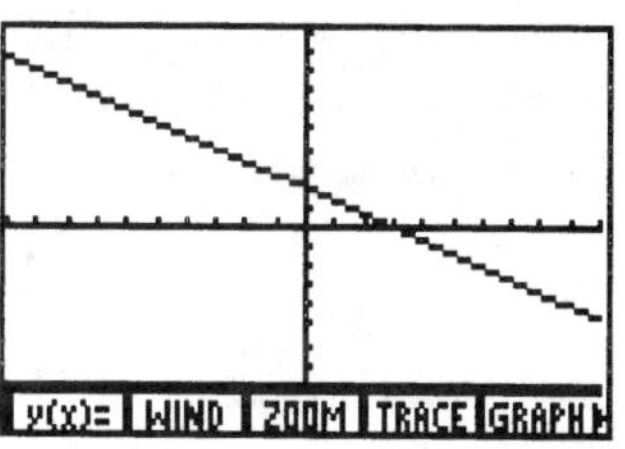

Decide which side of the graph is the solution region. Select a test point, say $(0,0)$, and see if it makes the inequality true. You can use the TEST feature of the TI-86 to do this.

Exit to the home screen, and press CLEAR to obtain a cleared screen. Since $x = 0$ and $y = 0$, press 2 (0) + 3 (0) 2nd [TEST] F4 ($\leq$) 6 ENTER. The calculator returns the value 1.

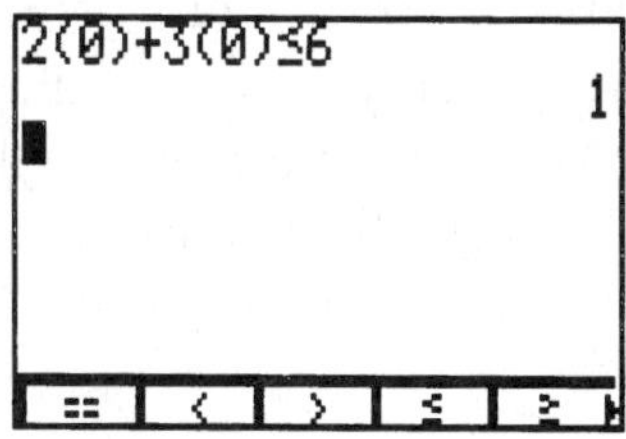

A value of 1 from this test indicates that the point (0, 0) makes the inequality true (satisfies the inequality). Thus, the set of points in the half-plane below the graph, including those points on the graph, make up the solution set.

Since you are expected to shade the part you *do not* want, you will have to shade the region *above* the graph.

Press [GRAPH] [F1] to access the equation editor. Press [MORE] to view the STYLE menu option. With the cursor at y1, press [F3] (STYLE) twice to select the graph style as "shade above". This will option will be indicated by the symbol ▜ appearing to the left of y1.

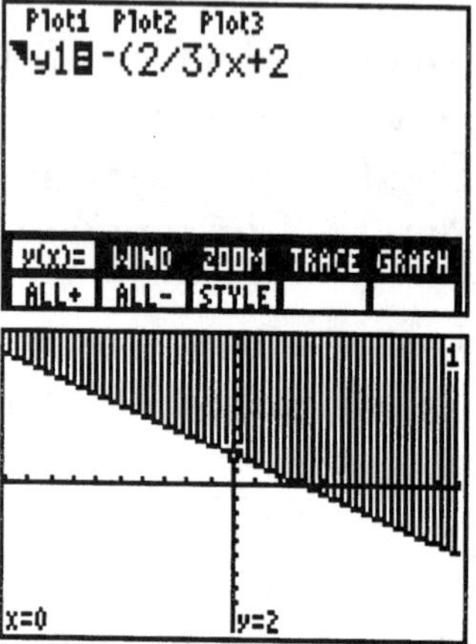

Graph this expression with the selected graph style by pressing [EXIT] [F5]. Press [F4] (TRACE) to obtain the graph screen shown at the right.

If you did not obtain a graph, go back into the equation editor and make sure that the = sign is highlighted. If not, then the equation is said to be deselected (which simply means "not selected to be graphed"). If this is the case, then press [F5] (SELCT) and then [2nd] [F5] (GRAPH) to obtain the graph.

Inequalities and Vertical lines
Since vertical lines are not functions, they cannot be graphed using the equation editor. However, vertical lines can be drawn on the graph screen. This is done with the vertical line feature VERT found in the DRAW menu of the graph editor [GRAPH].

To see this, access the equation editor ([GRAPH] [F1]), deselect all equations (with the cursor on each equation line press [F5]), and press [EXIT] [MORE] [F2] (DRAW) [F3] (VERT) and you will be taken to the graph screen.

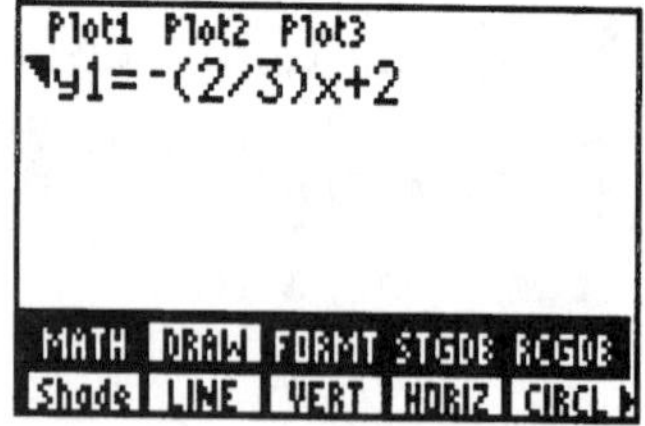

Move the left or right arrow keys to see the vertical line move across the screen. At or near an x value of your choice, press [ENTER] to establish a vertical line there. Continue in this fashion to establish other vertical lines as desired. Press [CLEAR] when finished drawing all of the vertical lines you desire.

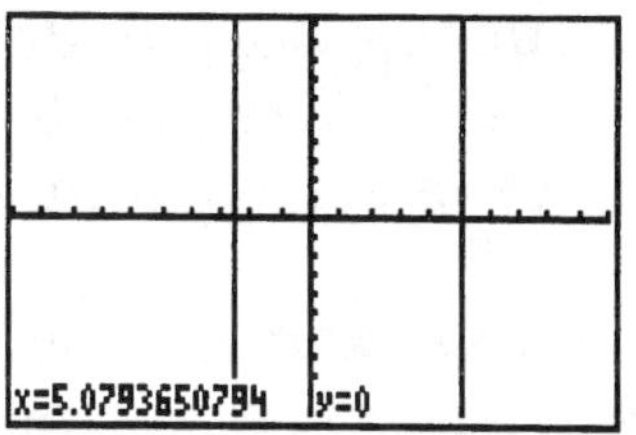

Vertical lines are drawings and not graphs on the TI-86 (in the function mode). Therefore no calculations can be performed on them. Also, these lines will remain on the graph screen until you either (1) clear them using the "clear draw" command (CLDRW) in the DRAW menu ([GRAPH] [MORE] [F2] (DRAW) [MORE] [MORE] [F1] (CLDRW)), or (2) by changing the graph window settings. The latter option is frequently used.

Graphing inequalities such as $x \geq 4$ can also be done with this calculator. Set the graph window to the standard viewing window. From a clear home screen, go to the DRAW menu in the graph editor ([GRAPH] [MORE] [F2] (DRAW)) and select SHADE ([F1]). The Shade(command should be pasted on your home screen.

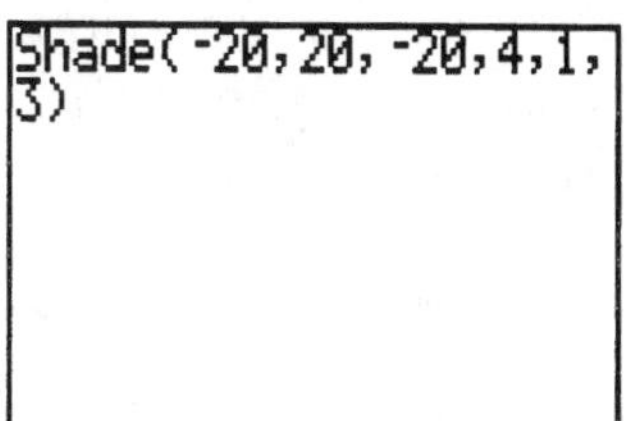

At the prompt, enter [(-)] 20 [,] 20 [,] [(-)] 20 [,] 4 [,] 1 [,] 3 [)] [ENTER] to see the solution. The command issued does the following: graphs from $y = -20$ to $y = 20$, shading between $x = -20$ to $x = 4$. The 1 says to use a vertical pattern for the shading, and the 3 says to space the pattern out every third pixel (addresses the quality of the shading).

Remember, *the part you do not want will be shaded, leaving the solution region blank.* Thus the solution region is the region to the right of the shading, since these numbers are greater than or equal to 4. As discussed above with vertical lines, this is not a graph but a drawing; therefore no calculations can be performed. The drawing can be cleared in the same manner as listed for the vertical lines in the previous paragraph.

Finding Corner Points

Let's use Example 3 *Corner points* in the text to demonstrate how to graph a
system of inequalities as well as how to find corner points using the TI-86.

In this example, you were asked to sketch the solution region of the system
$$3x - 2y \le 6$$
$$x + y \ge -5$$
$$y \le 4$$
and to find the coordinates of all the corner points.

To begin, solve each inequality for y and input the
three corresponding equations into the equation
editor. Select "line" for the graph style for each
equation. This style is indicated by the symbol ╲ in
front of the equation. With the graph viewing
window at the standard setting, graph these
equations.

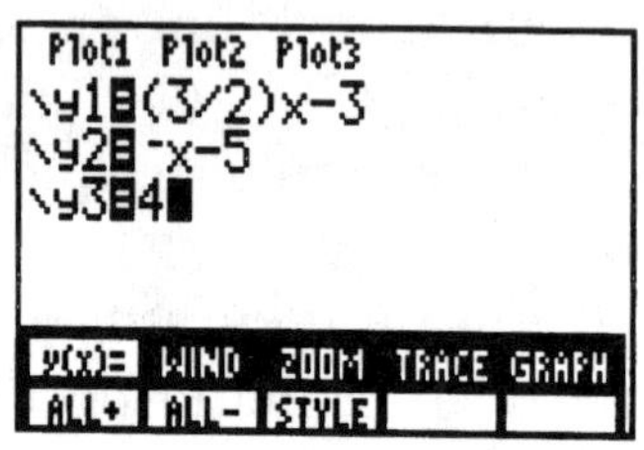

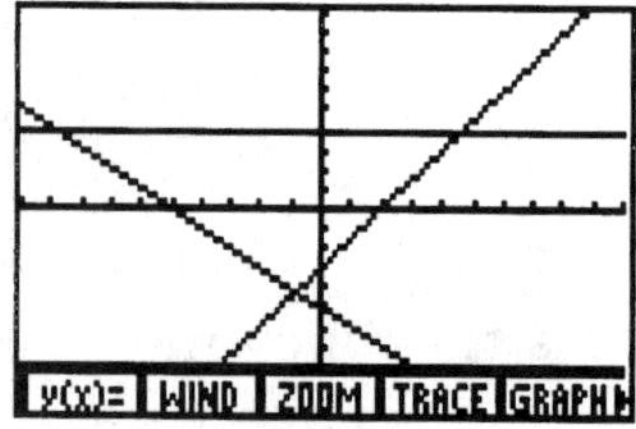

Starting from the top of the screen and moving left,
you can count seven different regions created by these
three graphs, including the triangular center region.
You must decide which of these will be the solution
region. What's a good place to start checking? The
origin, of course!

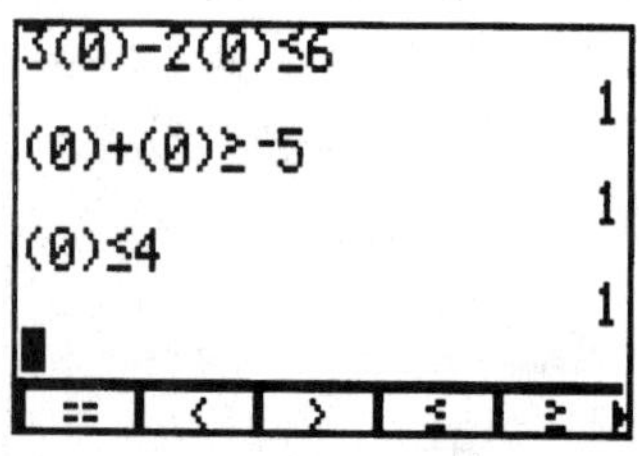

Using (0,0) as a test point, use the TEST feature to
determine if this point will make each of the three
inequalities in the system true. On a clear home
screen, press [2nd] [TEST] to access the TEST menu, and
then use the appropriate inequality symbols to test
each inequality as shown in the screen to the left.

Since 1 was returned for each test, the point (0,0) makes each inequality true.
The solution region is therefore the center region that
includes this point. Thus, *you must shade all other
regions.*

Return to the graph screen, press [F4] (TRACE) and use
the up and down arrow keys to toggle between the

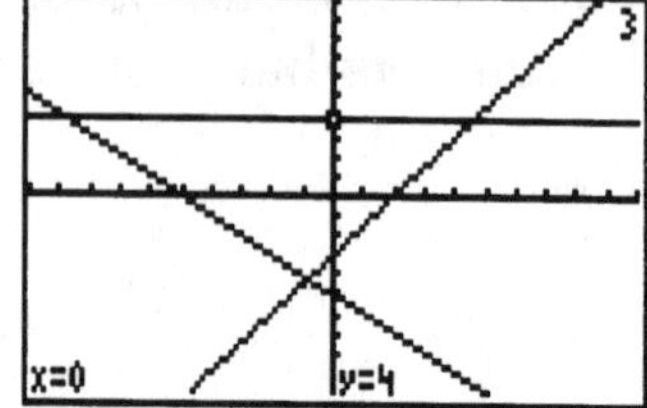

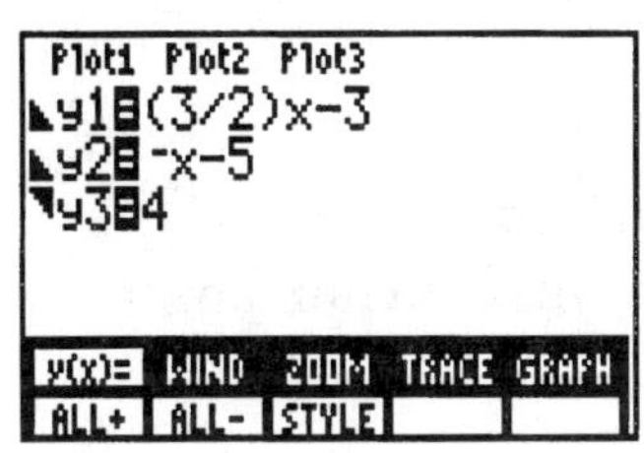

graphs and determine which graph corresponds to which equation (look at the numbers in the upper right corner of the graph screen). Having done so, you will notice that you have to shade below both y1 and y2 and above y3. Press [GRAPH] [F1] and change the graph style for each equation accordingly.

Once you have done this, graph your selections and observe the solution region as being the non-shaded (white) region. In applications, this region is referred to as the *feasible region*.

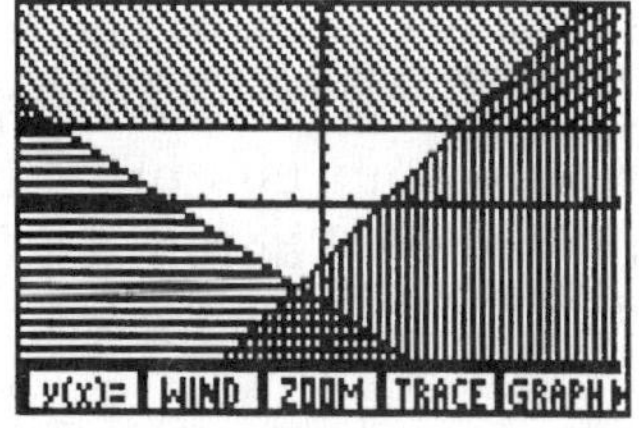

To find the corner points, press [MORE] [F1] (MATH) [MORE] to access the intersection command ISECT. Point A, the left most point as identified in the text, is the point intersected by the graphs of y2 and y3.

To find this point, press [F3] (ISECT) to obtain the prompt First curve?. Paying attention to the number in the upper right corner of the screen, press [▼] to place the cursor on y2 and press [ENTER]. Press [ENTER] again to choose y3 for the Second curve? selection. Press [ENTER] again and the coordinates of the corner point will appear as the intersection.

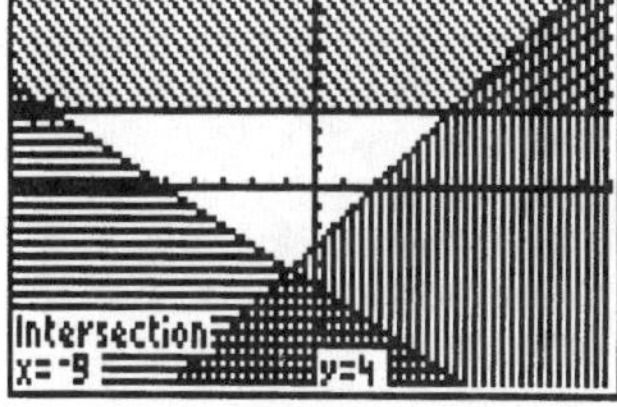

To find point B, the right most point, note that it is intersected by the graphs of y1 and y3. Press [GRAPH] [MORE] [F1] [MORE] [F3] to invoke the ISECT command. Press [ENTER] to choose y1 as the first curve. Press [▼] twice to place the cursor on y3, and press [ENTER] to choose this graph for the second curve. Finally, press [ENTER] at the Guess? prompt and the coordinates will appear on the screen.

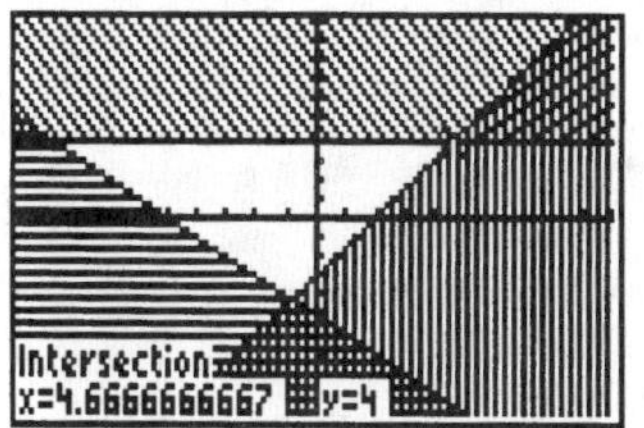

Repeat this procedure for the last corner point, point C. Note that the cursor can sometimes be difficult to see in the midst of the shading, so use your arrow keys leisurely to verify that you are selecting the correct curves for the appropriate corner point. Your result should be the same as shown in the screen at

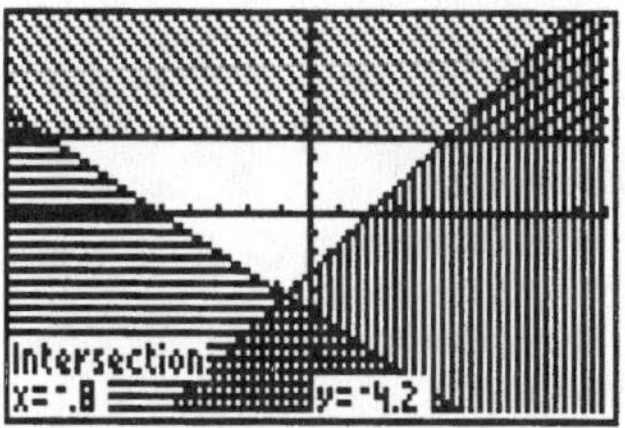

the right of this paragraph.

When the intersection is found, the x and y values displayed are automatically stored in memory. For example, when the last corner point was found, $-.8$ was stored as x and -4.2 was stored as y.

To obtain the fractional representation of these numbers, exit to the home screen, and press [x-VAR] [CUSTOM] [F2] (▶FRAC) [ENTER] (you were shown how to set up your custom menu in section 1.3 of this manual). Then press [2nd] [alpha] [y] [F2] (▶FRAC) [ENTER]. This gives you the coordinates of the point as given in the text, namely $\left(-\dfrac{4}{5}, -\dfrac{21}{5}\right)$!

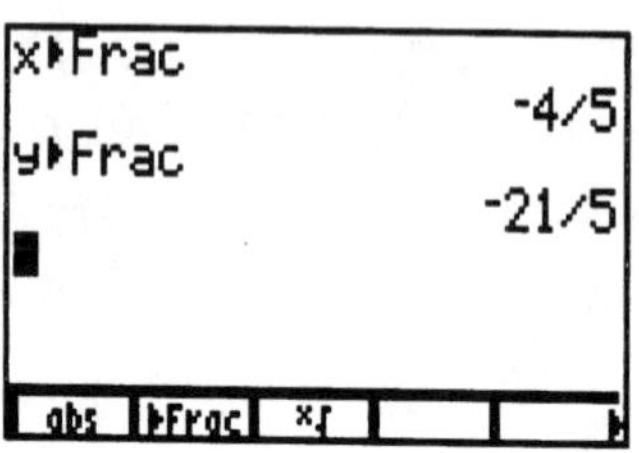

Find the second corner point again and try this for the x-value 4.6666666667. You should obtain $\dfrac{14}{3}$.

4.2 Solving Linear Programming Problems Graphically

In this section you are given a procedure for using the TI-86 to solve linear programming problems. In the text, Example 6 shows you how to use Excel to solve a linear programming problem. The solver mentioned in the example with Excel is not available on the TI-86, but by applying what you have learned in the previous section of this manual, you can use your graphing calculator technology to solve this problem.

The Resource Allocation problem of Example 6 (and Example 5) is the following:

$$
\begin{aligned}
\text{Minimize} \quad & c = 200x + 400y && \text{Objective function} \\
\text{Subject to} \quad & 2x + y \geq 200 && \text{Constraint 1} \\
& x + 3y \geq 300 && \text{Constraint 2} \\
& 3x - y \geq 0 && \text{Constraint 3} \\
& x \geq 0, \; y \geq 0 && \text{Constraint 4, Constraint 5}
\end{aligned}
$$

Recall that the Graphical Method for Solving Linear Programming Problems in Two Unknowns was given in text as
1. Graph the feasible region.
2. Compute the coordinates of the corner points.
3. Substitute the coordinates of the corner points into the objective function to see which gives the optimal value.

We now give the procedure to carry out this method using the TI-86. For each instruction, the general instruction will appear first. Then any instructions that are specific to Example 6 will be given in the succeeding paragraph of the corresponding step.

1. Solve each of the constraint inequalities for y and input the corresponding equations into the equation editor.

2. For each equation you entered from step 1, decide which side of the graph to shade (the side that is *not* the solution for the individual inequality). Choose the graph style representing your decision.

Graph the equations, and use the graph to determine a point in which to test each inequality. When selecting a test point for this problem, you can not use the origin (0, 0) because one of the equations ($y = 3x$) goes through this point.

Choosing the point (100, 125), you can use the TEST feature of your calculator to see that the region containing this point is the feasible region. You can now set the shading styles on your equations in the equation editor.

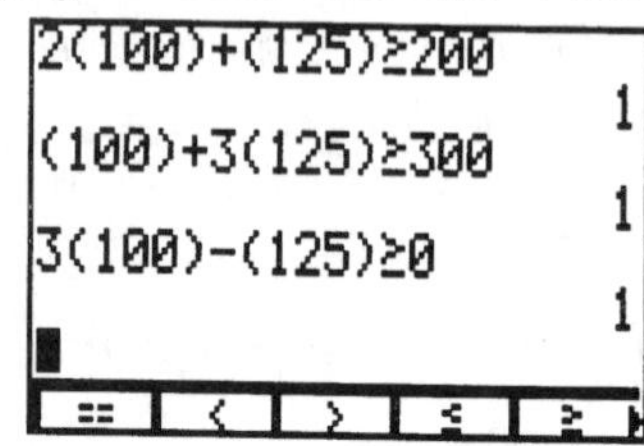

3. Set the graph viewing window. Use the intercepts as a guide.

For this problem, set the viewing window to $0 \leq x \leq 350$ (xScl = 50) by $0 \leq y \leq 250$ (yScl = 50). By setting xMin and yMin to 0, you are addressing and satisfying Constraints 4 and 5.

4. Graph the feasible region.

Note that for this example, there are three corner points, as you can not overlook the point on the x-axis that is hidden by the menu. Press [CLEAR] to view the graph without the menu shown. Press [GRAPH] to retrieve the menu.

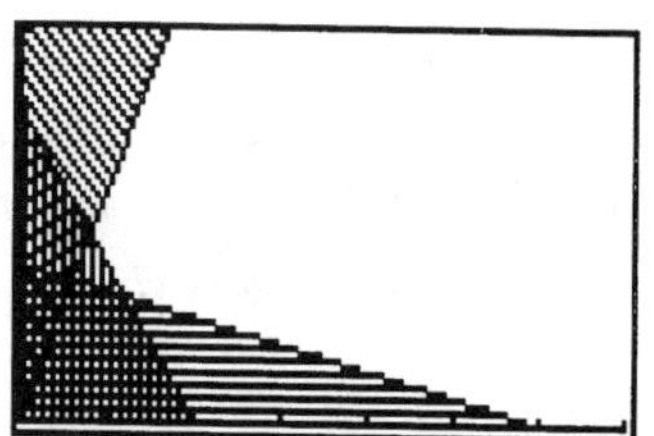

5. Use ISECT and ROOT from the MATH menu of the graph editor to find the corner points. Keep track of them by writing them down as you find them.

In this example, one of the corner points is the x-intercept, so you will need to use the ROOT command to identify this point instead of ISECT.

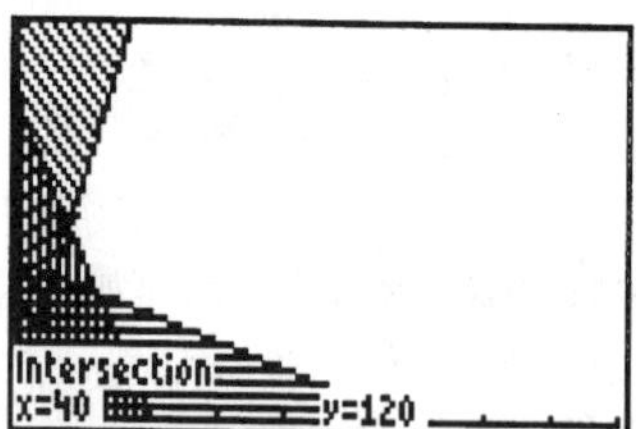

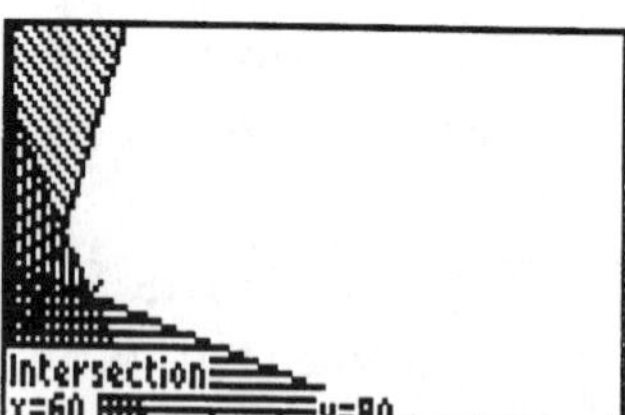

Find the first two corner points using the ISECT command. The results are shown above.

To find the third corner point, select ROOT from the GRAPH MATH menu and press ⊡ to place the cursor on the graph of y2=(300-x)/3. For a Left Bound, arrow near the intercept and press ENTER anywhere to the

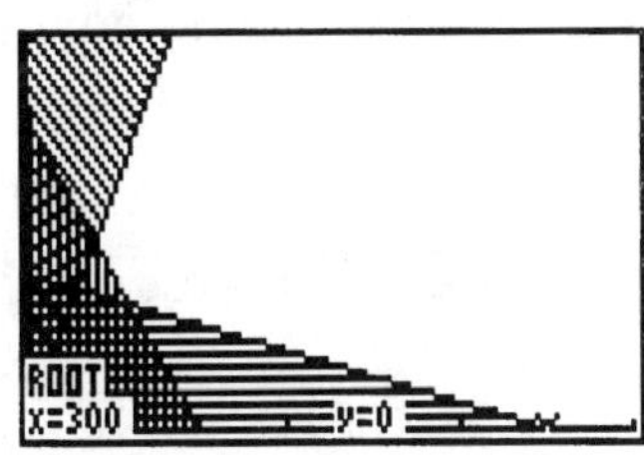

left of it (the *y* value should be positive). Then, arrow to the right of the intercept (the *y* values should be negative) and press ENTER to select a Right Bound. Arrow near the intercept, between the Left Bound and Right Bound, and press ENTER for the Guess. The *x*-intercept will appear on he screen labeled as ROOT.

Note that a little algebra (let $y = 0$), paper, and pencil would probably have been a quicker way to obtain this corner point!

6. If the feasible region is unbounded, create a rectangle around the region that includes the corner points.

Since the feasible region in this example is unbounded, create a bounded region by adding a convenient rectangle that encloses all three corner points. This rectangle can be created by choosing the lines $y = 150$ and $x = 400$.

Redefine the graph viewing window for *x* to be $0 \leq x \leq 450$. Input $y = 150$ into the equation editor as ▜y4=150, noting the graph style selected (to shade away from the feasible region). Use VERT in the GRAPH DRAW menu to create the vertical line $x = 400$.

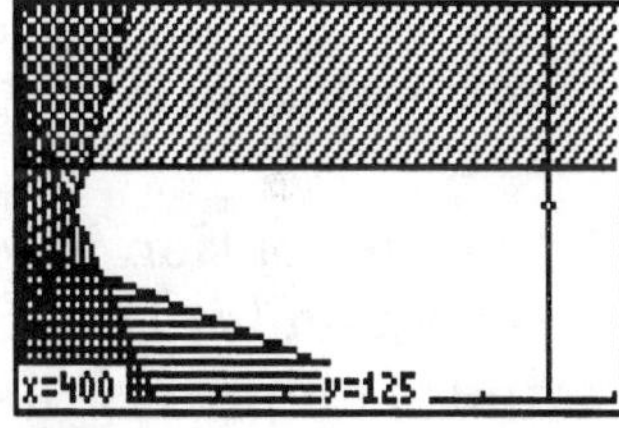

In some instances, the vertical line may not land directly on the value you desire, so set it near the value and remember what the value you intended it to be when identifying the new corner points. Also, unless you use the SHADE feature, you must remember that the region to the right of the vertical line is part of the *non-solution* region.

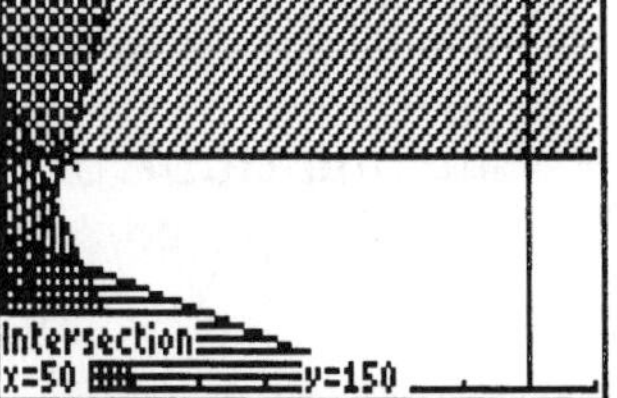

Now identify the three new corner points of the bounded region. One of the points is shown in the accompanying screen (above). The second point is where the lines $x = 400$ and $y = 150$ intersect which is (400, 150), and the third point is where the vertical line intersects with the x-axis, which is (400,0).

7. Exit to the home screen and evaluate the objective function at the corner points.

Using the corner points (40,120), (60,80), (300,0), (50,150), (400,150), and (400,0), evaluate the objective function on the (cleared) home screen. One way is to do the following.

Press 40 [STO▸] [x-VAR] [2nd] [:] [ALPHA] 120 [STO▸] [2nd] [alpha] [y] [2nd] [:] [ALPHA] [ALPHA]. This defines x as 40 and y as 120 and returns the original cursor. Now enter the objective function: press 200 [x-VAR] [+] 400 [2nd] [alpha] [y]. Press [ENTER] to evaluate the objective function at the point (40,120).

```
40→x:120→y:200 x+400y
                 56000
60→x:80→y:200x+400y
                 44000
300→x:0→y:200x+400y
                 60000
50→x:150→y:200x+400y
                 70000
400→x:150→y:200x+400y
                140000
400→x:0→y:200x+400y
                 80000
```

Press [2nd] [ENTER] to recall this command, change 40 to 60 and 120 to 80 (using [DEL] as necessary), and press [ENTER] to evaluate the objective function at (60,80). Continue this process, using [DEL] (delete) and [2nd] [INS] (insert) as necessary, to evaluate the objective function at the other corner points. Be sure to record your results in order to make a decision on the optimal solution.

Another option is to store the lists of corresponding x and y coordinates as shown, enter the objective function 200x+400y, and then press [ENTER] to evaluate the objective function. Use [▶] and [◀] to scroll and view the entire list of possible optimal values. The optimal (minimum) value 44,000 indicates that the optimal solution is found at the corner point (60,80).

```
{40,60,300,50,400,400
}→x
{40 60 300 50 400 40...
{120,80,0,150,150,0}→
y
   {120 80 0 150 150 0}
200 x+400y
{56000 44000 60000 7...
```

Can you think of another way using y5 in the equation editor?

4.3 The Simplex Method: Solving Standard Maximization Problems

In this section you will see how to use the TI-86 to perform the simplex method for solving standard maximization problems in n unknowns. The pivoting technique presented in Chapter 2 will be needed. Recall that the pivot element is used to make all the entries in the pivot column into 0 (except the pivot element). We will use Example 4 *Using Pivoting Technology* from the text to illustrate use of the calculator to perform the simplex method.

However, before we do so, it may be a good idea to review the calculator's matrix row operation commands presented in Chapter 2.

Matrix row operations on the TI-86

The row operation commands are found in the `MATRX` editor by pressing `2nd` `[MATRX]` `F4` (`OPS`) `MORE`. The syntax for using the operation commands are given below.

`rSwap(`
- **Syntax:** `rSwap(matrix name,row i,row j)`
- **Purpose:** Interchange (swaps) row i with row j
  - **EXAMPLE: RSWAP(A,2,3)→B SWAPS ROW 2 AND ROW 3 OF MATRIX A AND STORES THE RESULT AS MATRIX B.**

`rAdd(`
- **Syntax:** `rAdd(matrix name,row i,row j)`
- **Purpose: Adds row i to row j and places the result in row j**
- **Example: `rAdd(A,1,3)→B` takes row 1 and row 3 of matrix A, adds them together, places the answer in row 3, and then stores the new matrix as matrix B.**

`multR(`
- **Syntax:** `multR(constant multiplier,matrix name,row i)`
- **Purpose: Multiplies row i by a constant and then replaces row i with the result**

- Example: `multR(-4,A,3)→B` multiplies row 3 of matrix A by –4, replaces row 3 with the answer, and then stores the new matrix as matrix B.

mRAdd(

- Syntax: `mRAdd(constant multiplier,matrix name,row i,row j)`
- Purpose: Multiplies row i by a constant, adds the result to row j, and places row j with the answer
- Example: `mRAdd(2/3,A,2,1)` multiplies row 2 of matrix A by 2/3, adds the result to row 1, replaces row 1 with the answer to this addition, and then stores the new matrix as matrix B.

Remember that in the text, the row operations are written as $aR_{change} \pm bR_{pivot}$ whereas the calculator's row operation commands require *pivot row first* and *change row last*. Also, it is important to store the results from each row operation performed as another matrix to keep from having to return to the beginning if a mistake occurs.

Example 4 *Using Pivoting Technology*

Maximize $\quad p = 1.3x + 4.5y + 11.3z$

subject to
$$2.4x - 0.21y + 11.7z \le 5$$
$$3.3x + 5.6y - 2.35z \le 5$$
$$x - y + 2.3z \le 10$$
$$x \ge 0, y \ge 0, z \ge 0$$

Enter the initial tableau as matrix A ([2nd] [MATRX] (EDIT) [A] [ENTER]). Enter the dimension as 4×8 and then enter the elements of the tableau. Afterwards, exit to the home screen and view the tableau ([ALPHA] [A] [ENTER]) to verify that you have correctly input all entries. This is how the matrix should look over two screens.

$$[[2.4 \quad -.21 \quad 11.7 \quad 1 \quad 0 \quad 0 \quad 0 \quad 3\,]$$
$$[3.3 \quad 5.6 \quad -2.35 \quad 0 \quad 1 \quad 0 \quad 0 \quad 5\,]$$
$$[1 \quad -1 \quad 2.3 \quad 0 \quad 0 \quad 1 \quad 0 \quad 10]$$
$$[-1.3 \quad -4.5 \quad -11.3 \quad 0 \quad 0 \quad 0 \quad 1 \quad 0\,]]$$

Recognizing that the most negative entry in the bottom row is -11.3 (row 3) and that the smallest test ratio is $3/11.7 = 0.2564$, we conclude that the pivot is 11.7, the 1,3-entry.

Convert this pivot element to 1 by dividing row 1 by 11.7. Pull up the row operations in the MATRX OPS menu ([2nd] [MATRX] [F4] [MORE]) and enter the command on the multR(1/11.7,A,1)→B [ENTER] on the home screen. The result (shown by scrolling the entire matrix):

We suggest recording one column at a time, being sure to check for accuracy. Also, you should round as necessary on your paper.

Now you must use the pivot element to zero out the remaining entries in column three. To do so, you need the operations $R_2 + 2.35R_1$, $R_3 - 2.3R_1$, and $R_4 - 11.3R_1$.

You can do this one row at a time, viewing the result each time, or via a single command line using the colon [:]. Do this by entering the command mRAdd(2.35,B,1,2)→C:mRAdd(-2.3,C,1,3)→D:mRAdd(11.3,D,1,4)→E on the home screen.

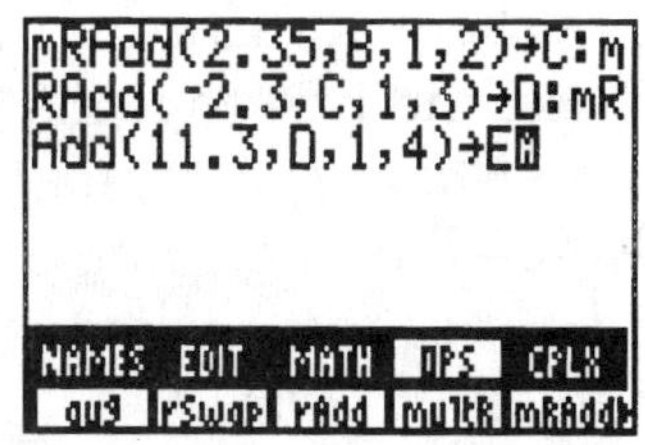

You must store the resultant matrix from each command as a new matrix in order to update the tableau with its zeroed out entry. The new tableau is shown below over several screens.

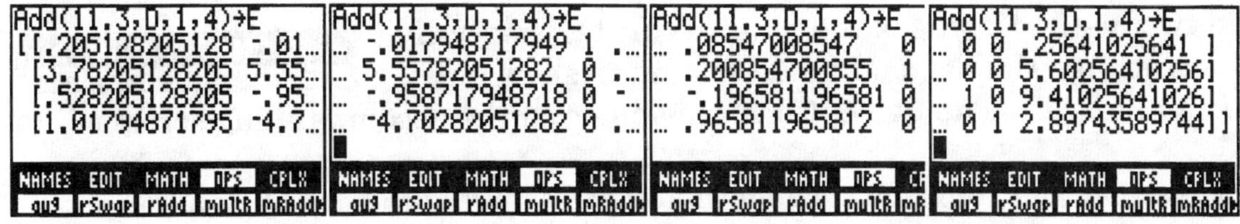

Repeat the process, determining that the pivot is now 5.55782, the entry in
the 2,2 position. Turn the pivot into 1 with the command
`multR(1/5.55782051282,E,2)→F`.

Recall from section 3.1 of this manual (*Referring to
and Editing the Entries of a Matrix*) that we can
reference a specific matrix entry using the command
matrix name(`row,column`). Thus `E(2,2)` will
reference the 2,2-entry of matrix E (the pivot
element). You can use this fact to turn the pivot into
1 with the command `multR(1/E(2,2),E,2)→F`.

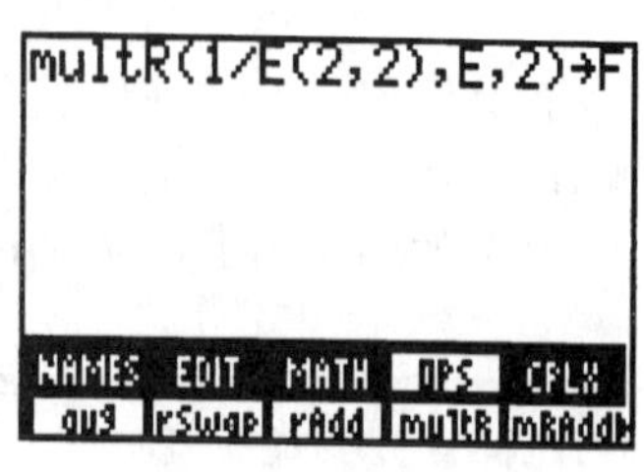

$$\begin{bmatrix} .205128205128 & -.017948717949 & 1 & 0 & 0 & 0 & .25641025641 \\ .680491799497 & 1 & 0 & .036139112516 & .179926645291 & 0 & 0 & 1.008050564 \\ .528205128205 & -.958717948718 & 0 & -.196581196581 & 0 & 1 & 0 & 9.41025641026 \\ 1.01794871795 & -4.70282051282 & 0 & .965811965812 & 0 & 0 & 1 & 2.89743589744 \end{bmatrix}$$

These screens are getting (possibly *have been*) rather difficult to read, so if
you wish, you could change the decimal mode to 2 (2nd [MODE] ⊡ ▶ ▶ ▶ ENTER
EXIT), but we will not do so.

Use the pivot to zero out the rest of the column. Again, you can use a single
command line to issue multiple commands. Also, you
can either type in the values needed (to five or six
decimal places) from column two, or you can
reference the needed entry using *matrix
name*(`row,column`).

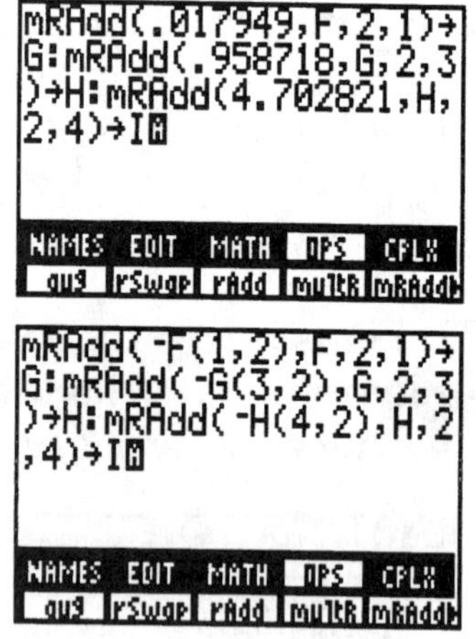

Obviously, the second option is more challenging as
you really have to think about which entry to
reference, which matrix to reference it from, and how
to use the pivot value to zero out the entry. But of
the two methods, this is the most accurate, especially
if you change the decimal mode setting to show only 2
decimal places. The two options are shown in the accompanying screens.

The result from the reference method (the second option) is shown below.

```
,4)→I                         ,4)→I                ,4)→I                  ,4)→I
[[.217342160504 0 1 …   ….086118736208    … .003229452608  … 0 0 .274503471662]
 [.680491799497 1 0 …   ….036139112516    … .179926645291  … 0 0 1.008050564  ]
 [1.18060483034 0 0 …   …-.161933980762   … .172498904293  … 1 0 10.3766925792]
 [4.21817951143 0 0 …   …1.13576772547    … .846162718276  … 0 1 7.63811676778]]
 NAMES EDIT MATH OPS CPLX   NAMES EDIT MATH OPS  NAMES EDIT MATH OPS  NAMES EDIT MATH OPS CPLX
 gu9 rSwap rAdd multR mRAdd  gu9 rSwap rAdd mult  gu9 rSwap rAdd mult  gu9 rSwap rAdd multR mRAdd
```

Because there are no negative numbers in the bottom row, you are finished.
Refer to the text to see how to read the solution.

4.4   The Simplex Method: Solving General Linear Programming Problems

There are no new calculator techniques introduced in this section.

4.5   The Simplex Method and Duality

There are no new calculator techniques introduced in this section.

You're the Expert – Airline Scheduling

There are no new calculator techniques introduced in this section.

CHAPTER 5

THE MATHEMATICS OF FINANCE

TI-86

```
N=60
I%=8.5
PV=19000
·PMT=-389.814095214
FV=0

 TVM  FUNC  VARS FORMT SOLVE
```

5.1   Simple Interest

This section has no new calculator skills. All computations can be done on
the home screen.

5.2   Compound Interest

Compound interest problems can be calculated on the home screen. Unlike
the TI-83 mentioned in the text and the TI-83 Plus discussed in this manual,
the TI-86 does not have a built-in Finance feature. However, some options do
exit, and we will discuss both of them in this section and the next.

The first option is to use the Time Value of Money program TVMPr written
by the authors of this manual and presented herein. You will have to take
the time to input this program into your program editor. The program is
listed at the end of this section. Refer to pages 214-220 of your TI-86
Graphing Calculator Guidebook you received with your calculator for
assistance with entering programs.

The second option is to download the Finance package of programs from the
Texas Instruments web site (www.ti.com). There are distinct advantages to
this option. First, the packaged programs from TI consist of everything you
need that the TI-83 Plus has built in. Thus, this package will allow the TI-86
to perform the very same finance computations as the TI-83. Second, you will
be able to perform the computations illustrated in the text. Lastly, once you
load and activate the package, it will be part of your calculator as if it was
built in.

Ask your instructor if they have this Finance package that they can share
with you. If not, ask them if they are able to download it for you as you will
need a TI-Graph Link to download the files from the computer.

Loading the TI Finance Package

Once you have loaded the Finance package into your calculator, you still have to activate it to become a part of your calculator's features. To do this, follow these steps. The screens are shown after the last step.

1. Press [2nd] [CATLG-VARS] [F1] (CATLG) [LOG] to get to the top of the catalog list. Arrow down so that the selection cursor is pointing at Asm(.
2. Press [ENTER] to paste Asm(on the home screen.
3. Press [PRGM] [F1] (NAMES) to bring up the program names, and select Financ.
4. Press [ENTER] to obtain the "Caution" message shown below.
5. Press [F1] (Continue) and the installation is complete.

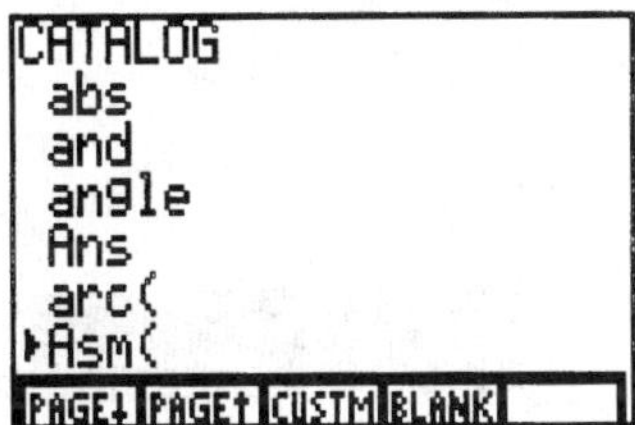 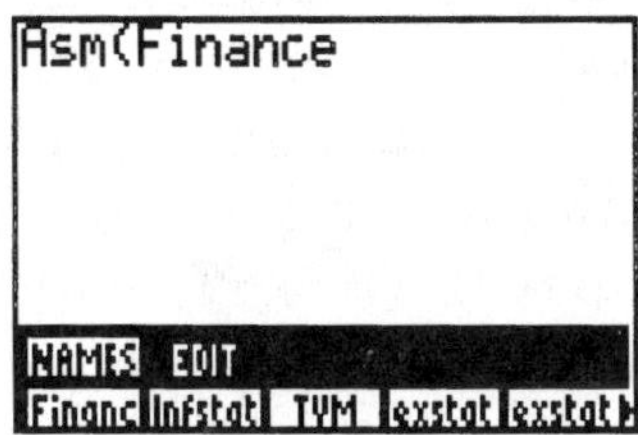 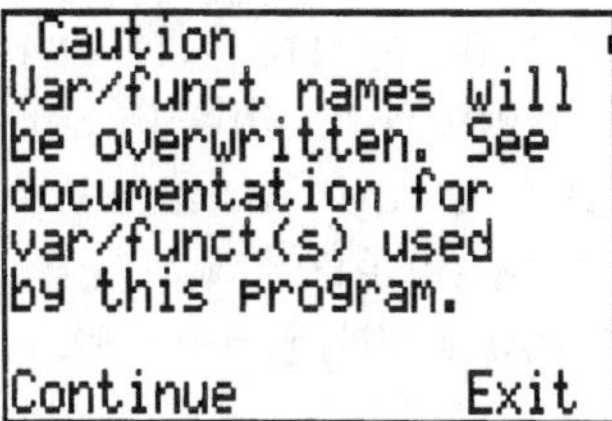

The Finance package has been installed as FIN under the MATH menu. To access it, press [2nd] [MATH] [MORE], and there it should be. You can also download 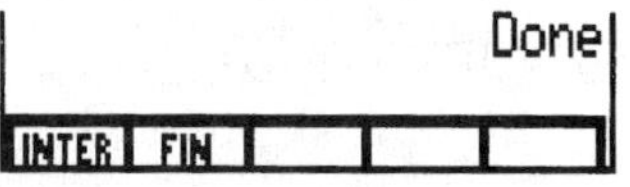 documentation referencing the use of this package from TI's web site. This does not require a TI Graph Link to do so.

Example 1 *Savings Accounts*

This example in the text is stated as follows. *In March, 1994, the Bank of Nova Scotia was paying 4.75% interest on savings accounts. If the interest is compounded quarterly, find the future value of a $2,000 deposit in six years. What is the total interest paid over the period?*

Method 1 – Computing on the Home Screen

To do this on the home screen, you should first write down the formula needed and then identify all of the parts of the formula that are provided in

the problem. Thus, for the formula $FV = PV\left(1 + \dfrac{r}{m}\right)^{mt}$, notice that $PV =$

$2,000, $r = .0475$, $m = 4$, and $t = 6$. Substitute these values into the formula to obtain $FV = 2000\left(1 + \dfrac{.0475}{4}\right)^{4\times 6}$.

Enter the expression `2000(1+0.0475/4)^(4*6)` on the home screen and press `ENTER` to obtain the result `2655.06270297`, making the first answer $2,655.06. The total interest paid was $2,655.06 - 2,000 = \$655.06$.

Method 2 – Using the TVMr Program

Press `PRGM` `F1` (NAMES) to see a list of the programs in your calculator. The PRGM NAMES menu shown here may look different from yours. In particular, I have loaded the Finance package from the TI web site. Assuming you have inputted the finance program from this manual into your calculator, you should have TVMPr (Time Value of Money Program) as one of your menu options.

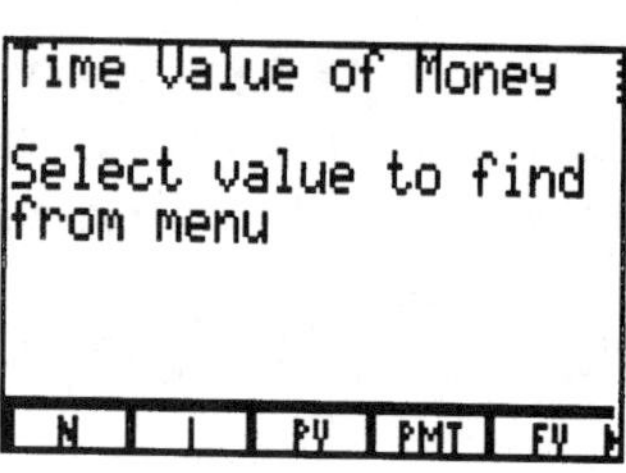

Press the function key selecting the program TVMPr and press `ENTER`. The screen given at the right shows what you should obtain. The ▶ shown at the very right of the menu options indicates that there is more to this menu. Press `MORE` to see the last menu item, which is QUIT. This allows you to quit the program.

Press `F5` (FV) for Future Value. At each prompt, enter the values known from the problem. *For the value you are trying to find, any number can be used, but you must give it a number.* I usually assign 0 to the desired variable.

PY is the number of payments per year, and CY is the number of compoundings per year. After entering the value for CY and pressing `ENTER`, you should obtain the screen shown at right, indicating the answer for the future value.

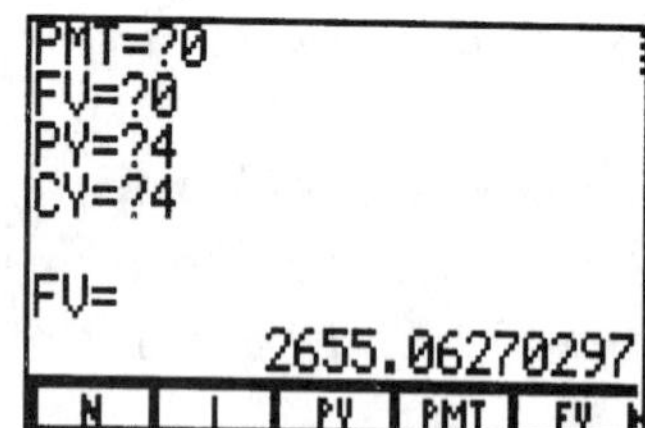

At this stage, you can press another function key to compute another value, or press `MORE` `F1` (QUIT) to stop the program. If you enter a value incorrectly and need to start the program over before finishing a computation, you can press $\boxed{\text{ON}}$ and an ERROR message will appear. Press $\boxed{\text{F5}}$ (QUIT) and then start over.

Method 3 – Using the TI Finance Feature

To use this feature, press $\boxed{\text{2nd}}$ [MATH] $\boxed{\text{MORE}}$ $\boxed{\text{F2}}$ (FIN) to access the Finance menu. Press $\boxed{\text{F1}}$ (TVM) to select the Time Value of Money option. Enter the values provided in the problem. The screens are shown below.

N=4*6	number of compounding periods over the 6 years
I%=4.75	actual interest rate, *not in decimal form*
PV=-2200	present value; negative since amount is given out
PMT=0	payment per period (no payment given in this section)
FV=0	future value; value we will compute so put 0

To complete the setup, press $\boxed{\text{F4}}$ (FORMAT) and enter the remaining values provided in the problem.

P/Y=4	payments per year
C/Y=4	number of times compounded per year
PMT:END BEGIN	not used in this section

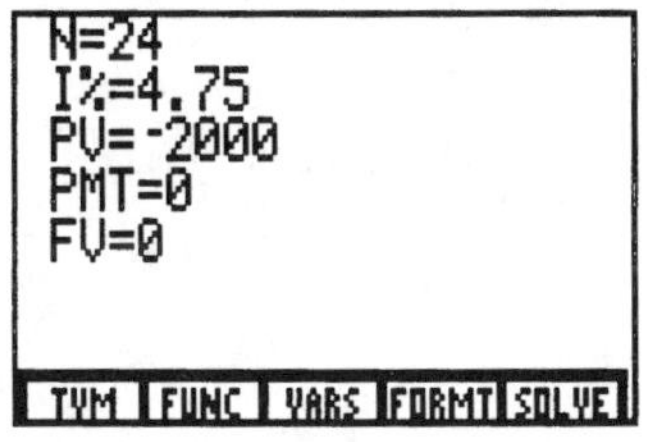
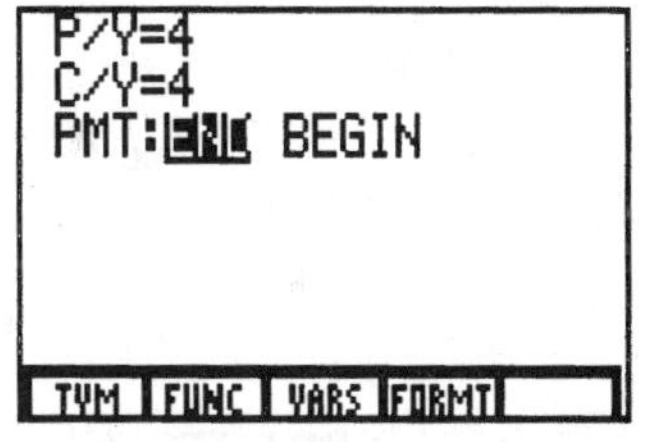

Finally, press $\boxed{\text{F1}}$ (TVM) again, arrow down so that the cursor is located at the variable you wish to solve for (FV in this case), and press $\boxed{\text{F5}}$ (SOLVE). The bullet next to FV indicates that the solution has been found for that variable.

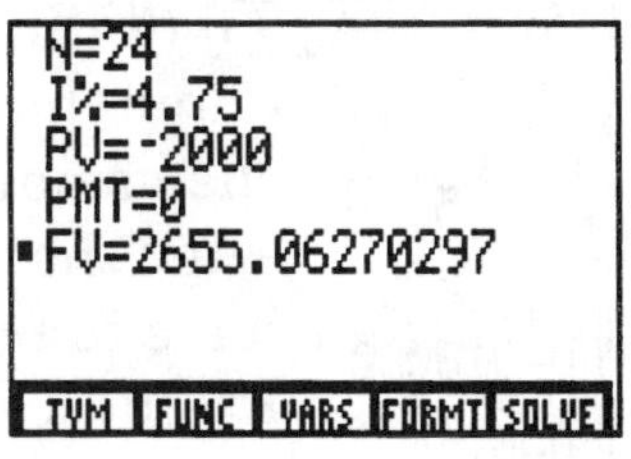

Once again, we see that the future value will be \$2,655.06. Regardless of the method used, the total interest paid over the period can be found on the home screen by entering 2655.06-2000 $\boxed{\text{ENTER}}$ to get 655.06.

Example 2 Zero-Coupon Bonds

The Megabucks Corporation is issuing 10-year zero-coupon bonds. How much would you pay for bonds with a maturity value of $10,000 if yu wish to get a return of 6.5% compounded annually?

This example requires you to find the present value *PV*. Again, we will look at the three methods we have to solve this problem: algebraically with the use of the home screen, using the program TVMPr, and using the TI Finance package.

Home screen

Solve the formula $FV = PV\left(1+\dfrac{r}{m}\right)^{mt}$ for *PV* to obtain $PV = \dfrac{FV}{\left(1+\dfrac{r}{m}\right)^{mt}}$. After identifying the values presented in the problem as $t = 10$, $FV = \$10{,}000$, $r = 6.5\%$, and $m = 1$, substitute these values into the formula for *PV* to get $PV = \dfrac{10{,}000}{\left(1+\dfrac{.065}{1}\right)^{10\times 1}}$. Enter this on the home screen as

`10000/(1+.065/1)^(10*1)` and press ENTER to see the result 5327.26035521.

Using Program TVMPr

Press PRGM F1 (NAMES), select TVMPr and press ENTER. Press F3 (PV) and enter the values provided in the problem. Remember, enter 0 for the variable you're trying to compute, namely PV. Pressing ENTER after the value for CY yields the result shown. Thus, you would have to pay $5,327.26 for the bond. Press MORE F1 (QUIT) to quit the program.

```
N=?10
I=?6.5
PV=?0
PMT=?0
FV=?10000
PY=?1
CY=?1
```

```
PV=
          -5327.26035521
 N   I   PV   PMT   FV
```

Using the TI Finance Feature

Press [2nd] [MATH] [MORE] [F2] (FIN) to access the Finance menu. Press [F1] (TVM) to select the Time Value of Money option. Enter the values provided in the problem.

N=10*1	number of compounding periods over the 10 years
I%=6.5	actual interest rate, *not in decimal form*
PV=0	present value; 0 (or any number) since we will compute it
PMT=0	payment per period (no payment given in this section)
FV=10000	future value; value of the bond at maturity

To complete the setup, press [F4] (FORMAT) and enter the remaining values provided in the problem.

P/Y=1	payments per year
C/Y=1	number of times compounded per year
PMT:END BEGIN	not used in this section

Finally, press [F1] (TVM) again, arrow down so that the cursor is located at the variable you wish to solve for (PV in this case), and press [F5] (SOLVE) to see the solution $PV = \$5,327.26$.

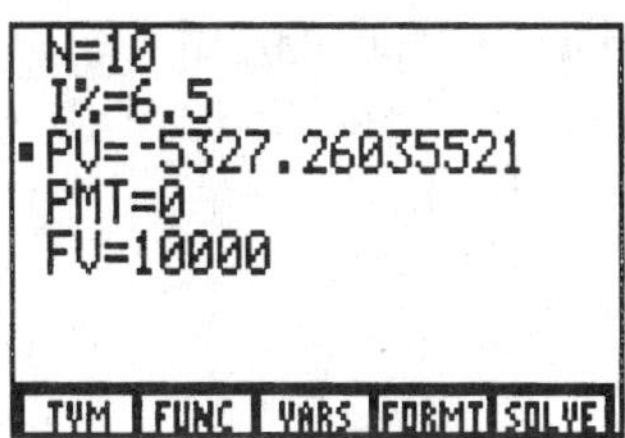

Example 3 *How Long to Invest*
You have \$5,000 to invest at 6% compounded monthly. How long will it take for your investment to grow to \$6,000?

Here you are asked to find the time t for your initial investment to grow to a certain amount. The same three methods are illustrated below.

Home Screen
Using logarithms, algebra, and the values provided in the problem ($PV = \$5,000$, $r = 0.06$, $m = 12$, and $FV = \$6,000$), solve the equation

$$FV = PV\left(1 + \frac{r}{m}\right)^{mt} \quad \text{for } t \text{ to obtain } t = \frac{\log 1.2}{12 \log\left(1 + \dfrac{0.06}{12}\right)}.$$

Enter on the home

screen the expression $\log\ 1.2/(12\ \log\ (1+0.06/12))$ and press [ENTER] to see the solution shown in the screen. Thus it would take about 3 years.

```
log 1.2/(12 log (1+.0
6/12))
                3.04628302993
```

Using the Program TVMPr

Press [PRGM] [F1] (NAMES), select TVMPr and press [ENTER]. Since you want to

```
CY=?12
N=
          36.5553963592
              Done
N/12
          3.04628302993
■
```

find the time t, press [F1] (N) and enter the values provided in the problem. Remember, enter 0 for the variable you're trying to compute, namely N. Pressing [ENTER] after the value for CY yields the result shown. Quit the program ([MORE] [F1] (QUIT)) and press [ALPHA] [N] [÷] 12 [ENTER] to find t.

Using the Finance Feature

Press [2nd] [MATH] [MORE] [F2] (FIN) to access the Finance menu. Press [F1] (TVM) to select the Time Value of Money solver. Enter the values provided in the problem.

N=0	number of compounding periods over the 10 years
I%=6	actual interest rate, *not in decimal form*
PV=-5000	present value; negative since you are giving this out
PMT=0	payment per period (no payment given in this section)
FV=6000	future value; value of the bond at maturity

Complete the setup by pressing [F4] (FORMAT) and enter the remaining values provided in the problem.

P/Y=12	payments per year
C/Y=12	number of times compounded per year
PMT:END BEGIN	not used in this section

Finally, press [F1] (TVM) again, and with the cursor at N press [F5] (SOLVE). After obtaining the screen shown here, exit to the home screen and divide N by 12 ([ALPHA] [N] [÷] 12 [ENTER]) to get the final answer.

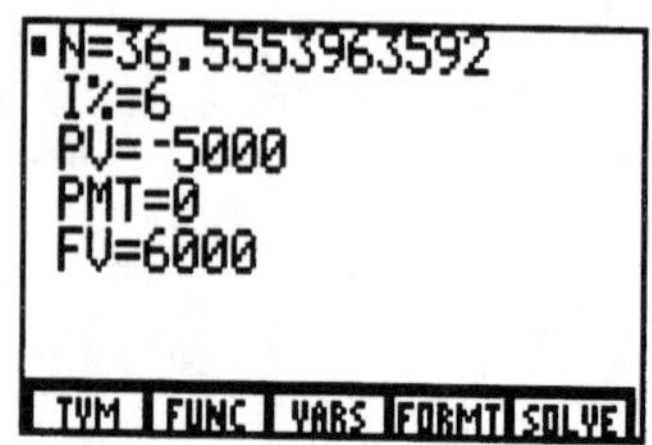

The Time Value of Money program TVMPr

Program TVMPr is given below. As stated at the beginning of the section,
you should refer to pages 214-220 of the TI-86 Graphing Calculator
Guidebook you received with your calculator for assistance with entering
programs.

Program TVMPr
```
:ClLCD
:Disp "Time Value of Money"
:Disp ""
:Disp "Select value to find"
:Disp "from menu"
:Lbl G
:0→J
:Menu(1,"N",A,2,"I",B,3,"PV",C,4,"PMT",D,5,"FV",E,6,"Quit",F)
:Lbl K
:Disp ""
:Prompt N
:Prompt I
:Prompt PV
:Prompt PMT
:Prompt FV
:Prompt PY
:Prompt CY
:1→J
:I/(100*CY)→i
:If H==1:Goto A
:If H==2:Goto B
:If H==3:Goto C
:If H==4:Goto D
:If H==5:Goto E
::Lbl A
:1→H
:If J==0
:Goto K
:-(-PMT+i*FV)/(i*PV+PMT)→P
:ln P/ln (1+i)→N
:Disp ""
:Disp "N=",N
:Goto G
:Lbl B
```

```
:2→H
:If J==0
:Goto K
:Solver(FV=-PV*(1+i)^N-PMT*((1+i)^N-1)/i,i,1)
:100i*CY→I
:Disp ""
:Disp "I=",I
:Goto G
::Lbl C
:3→H
:If J==0
:Goto K
:(i*FV-PMT*(1-(1+i)^N))/(i*(1+i)^N)→PV
:Disp ""
:Disp "PV=",-PV
:Goto G
::Lbl D
:4→H
:If J==0
:Goto K
:i(FV+PV(1+i)^N)/((1+i)^N-1)→PMT
:Disp ""
:Disp "PMT=",-PMT
:Goto G
::Lbl E
:5→H
:If J==0
:Goto K
:-PV*(1+i)^N-PMT*((1+i)^N-1)/i→FV
:Disp ""
:Disp "FV=",FV
:Goto G
::Lbl F
:Stop
```

5.3 Annuities, Loans, and Bonds

Annuities, loans, and bond problems can be calculated using the same three methods as mentioned in the previous section of this manual – algebraically with final computation on the home screen, using the program TVMPr, or using the Finance feature loaded into your calculator. Although you are encouraged to use either the program or the Finance feature, you will see all three methods, as you must take great care when entering expressions on the home screen.

This section also discusses how to build an amortization schedule. This can only be done on the TI-86 with the Finance feature.

Example 1 *Retirement Account*
Your retirement account has $5,000 in it and earns 5% interest per year compounded monthly. Every month for the next 10 years you will deposit $100 into the account. How much money will there be in the account at the end of those 10 years?

Here you are asked to compute the future value of an increasing annuity. Recognize that $PV = \$5,000$, $r = 5\%$ or 0.05, $m = 12$, $t = 10$, and $PMT = \$100$.

Home Screen
Using $i = r/m$ and $n = m \cdot t$, substitute the values into the formula

$$FV = PV(1+i)^n + PMT\left(\frac{(1+i)^n - 1}{i}\right) \text{ to get}$$

$$FV = 5000\left(1 + \frac{.05}{12}\right)^{120} + 100\left(\frac{\left(1 + \dfrac{.05}{12}\right)^{120} - 1}{\dfrac{.05}{12}}\right). \text{ On the}$$

home screen enter the expression on the right side of the last equation as
`5000(1+.05/12)^120+100(((1+.05/12)^120-1)/(.05/12))` and press ENTER to see the result.
You can also compute the two terms of the expression separately and then combine the results as done in the text.

```
5000(1+.05/12)^120+10
0((((1+.05/12)^120-1)/
(.05/12))
              23763.2754332
```

Using the Program TVMPr

Press PRGM F1 (NAMES), select TVMPr and press ENTER. Press F5 (FV) and enter the values provided in the problem. Remember, enter 0 for the variable you're trying to compute, namely FV. Pressing ENTER after the value for CY yields the result shown. Thus there will be $23,763.28 in the account after 10 years. Quit the program (MORE F1 (QUIT).

Using the Finance Feature

Press 2nd [MATH] MORE F2 (FIN) to access the Finance menu. Press F1 (TVM) to select the TVM solver. Enter the values provided in the problem.

N=120	number of compounding periods over the 10 years
I%=5	actual interest rate, *not in decimal form*
PV=-5000	present value; negative since paid this out already
PMT=-100	payment per period; negative since paying this out
FV=0	future value; use 0 since this value will be computed

Complete the setup by pressing F4 (FORMAT) and enter the remaining values provided in the problem.

P/Y=12	payments per year
C/Y=12	number of times compounded per year
PMT:END BEGIN	not used in this section

Finally, press F1 (TVM) again, and with the cursor at FV press F5 (SOLVE). The solution is found when the bullet appears next to the variable. The value in the account will therefore be $23,763.28. Press EXIT to return to the home screen.

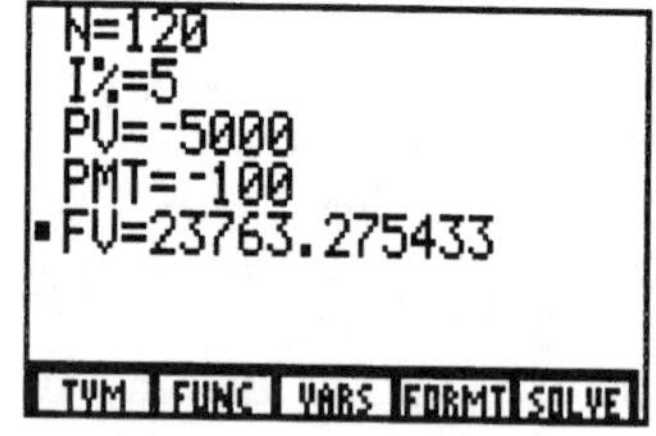

This is basically how all of the examples and problems in this section of the text are done. You are to identify the formula needed (especially if you're computing the value of the unknown on the home screen), identify all parts of the formula provided in the problem, then use the TVMPr program or the Finance feature to solve for the desired quantity.

Example 3 *Trust Fund*

You wish to establish a trust fund from which your niece can withdraw $2,000 every six months for 15 years, at which time she will receive the remaining money in the trust, which you would like to be $10,000. The trust will be invested at 7% per year compounded every six months. How large should the trust be?

Here you are asked to find the present value PV of a decreasing annuity with $FV = \$10,000$, $PMT = \$2,000$, $r = 7\%$ or 0.07, $m = 2$, and $t = 15$.

Home Screen

Using $i = r/m = 0.035$ and $n = m \cdot t = 30$, substitute the values provided into

the formula $PV = FV(1+i)^{-n} + PMT\left(\dfrac{1-(1+i)^{-n}}{i}\right)$ to get

$PV = 10000(1 + 0.035)^{-30} + 2000\left(\dfrac{1-(0.035)^{-30}}{0.035}\right)$. Enter

```
10000(1+.035)^-30+200
0((1-(1+.035)^-30)/.0
35
            40346.8749287
```

this on the home screen as `10000(1+.035)^-30+2000((1-(1+.035)^-30)/.035)` and press `ENTER` to see the result. Thus, the trust should be $40,346.87.

Using the Program TVMPr

Press `PRGM` `F1` (NAMES), select `TVMPr` and press `ENTER`. Press `F3` (PV) and enter the values provided in the problem. Enter 0 for the variable you're trying to compute, namely `PV`. Pressing `ENTER` after the value for `CY` yields the result shown. Again, you see that the trust should start with $40,346.87. Quit the program (`MORE` `F1` (QUIT).

```
N=?2*15
I=?7
PV=?0
PMT=?2000
FV=?10000
PY=?2
CY=?2
```

```
PV=
            -40346.8749287
  N    I    PV   PMT   FV
```

Using the Finance Feature

Press [2nd] [MATH] [MORE] [F2] (FIN) to access the Finance menu. Press [F1] (TVM) to
select the TVM solver. Enter the values provided in the problem.

N=30	number of compounding periods over the 15 years
I%=7	actual interest rate, *not in decimal form*
PV=0	present value; use 0 since this value will be computed
PMT=2000	payment per period; positive since she's receiving this
FV=10000	future value; amt to remain in the trust after the period

Press [F4] (FORMAT) and complete the setup.

P/Y=2	payments per year
C/Y=2	number of times compounded per year
PMT:END BEGIN	not used in this section

Now that all of the values provided in the problem
have been inputted into the calculator, press [F1]
(TVM), and with the cursor at PV press [F5] (SOLVE)
to obtain the solution.   The trust should start with
$40,346.87. Press [EXIT] to return to the home
screen.

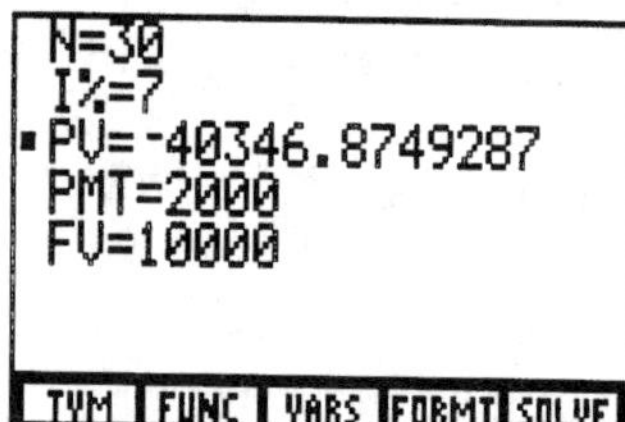

Note that the results for using TVMPr and the Finance feature differed in
sign from the result found using the formula and the home screen. Why?
What does it mean to have a negative value for the result using the program
and the Finance feature?

Amortization Schedule

An amortization schedule is a table indicating the details of repaying a loan.
This includes the number of the month of the payment, the outstanding
principal (principal amount remaining after the payment), the payment on
principal (amount of monthly payment that goes toward reducing the
principal), and the interest payment (amount of monthly payment that is
interest).

The Finance and table features of the TI-86 can be combined to create an
amortization schedule. Since the text is fairly clear on explaining how this is

done on the TI-83, and the procedure is somewhat similar on the TI-86, we will use a different example to illustrate using the TI-86 to develop the schedule.

You buy a car for a total cost of $20,500, of which you financed $15,800 at an interest rate of 8.5% for 5 years. Determine your monthly payment, and create an amortization schedule for the loan.

Here we have $PV = \$15,800$, $r = 8.5$, $t = 5$, $m = 12$, $FV = 0$, and $n = 12{\times}5 = 60$. Press [2nd] [MATH] [MORE] [F2] (FIN) to access the Finance menu. Press [F1] (TVM) to select the Time Value of Money option.

Enter the values provided in the problem. Place the cursor at PMT and press [F5] (SOLVE) to find the monthly payment. You will find that the monthly payment for your car will be $324.16.

Press [GRAPH] [F1] to access the equation editor. Press [2nd] [MATH] [MORE] [F2] (FIN) to access the Finance menu. Press [F2] (FUNC) [MORE] to access the functions bal, Σprn, and Σint. These functions and their syntax will be explained later.

To define y1 press [F3] (bal) [x-VAR] [,] 2 [)]. Press [ENTER] to go to y2. Define y2 by pressing [(-)] [F4] (Σprn) [x-VAR] [,] [x-VAR] [,] 2 [)]. Also define y3 as shown in the screen.

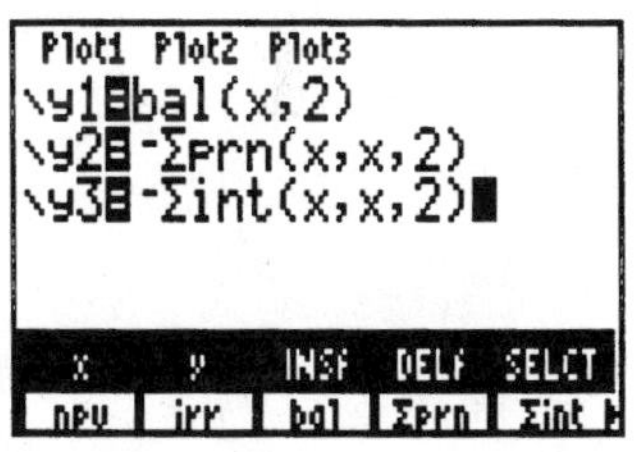

Press [TABLE] [F2] (TBLST) and put TblStart=0, ΔTbl=1, and set Indpnt: to Auto. Now press [F1] (TABLE) and you have the amortization schedule!

The second column, y1, is the outstanding principal (balance) after month x, y2 is the portion of the payment from month x that is applied to the principle (payment on principle), and third column, y3, is the interest payment from the months payment. You have to use the right arrow key to view y3.

The three Finance functions used to create the amortization schedule are given below.

bal(x,2) computes the principal balance at the end of month x, with the 2 indicating to round the result to 2 decimal places.

Σprn(x,x,2) computes the sum of the payments that goes toward principal from the end of month x to the end of month x, rounded to two decimal places. This gives the payment towards principal for the given month. Note that Σprn(m,n,2) computes the sum of the payments that goes toward principal from payment m through n. Can you explain why the (-) sign is used with this function for y2?

Σint(x,x,2) computes the sum of that part of the payment towards interest from the end of month x to the end of month x, rounded to two decimal places. This gives the portion of the payment towards interest for a given month. Note that Σint(m,n,2) computes the sum of the payments to interest from payment m through payment n. Why is this also accompanied by a (-) sign in y3?

To create an amortization schedule, you must always begin by using the Finance feature to compute one of the variables associated with the product you're attempting to amortize.

You're the Expert – Saving for College

There are not new calculator techniques presented here.

CHAPTER 6

SETS AND COUNTING

TI-86

6.1   Set Operations

This section requires no calculator skills.

6.2   Cardinality

This section has no new calculator skills. All computations can be performed on the home screen.

6.3   The Addition and Multiplication Principles

This section has no new calculator skills. All computations can be performed on the home screen.

6.4   Permutations and Combinations

In this section you will see how the TI-86 can be used to compute *permutations* and *combinations*. These will be useful in upcoming sections involving probability.

Permutations

The number of possible permutations of n items is given by $n!$ (read "n factorial"). This is defined as $n! = n \times (n-1) \times (n-2) \times \cdots \times 2 \times 1$. Thus, $5! = 5 \times 4 \times 3 \times 2 \times 1 = 120$ and $8! = 8 \times 7 \times 6 \times 5 \times 4 \times 3 \times 2 \times 1 = 40{,}320$.

The number of permutations of n items taken r at a time is given by the formula $P(n,r) = \dfrac{n!}{(n-r)!}$. Therefore the number of permutations of 10 items taken four at a time is

$$P(10,4) = \frac{10!}{(10-4)!} = \frac{10!}{6!} = \frac{10 \times 9 \times 8 \times 7 \times 6!}{6!} = 10 \times 9 \times 8 \times 7 = 5{,}040 \,.$$

The TI-86 can be used to compute the values in the preceding paragraphs directly. The features needed to do so are found in the math probability menu MATH PROB. To access this menu, press [2nd] [MATH] [F2] (PROB). Press 5 [F1] (!) [ENTER] to compute 5!. Compare the result to the one found in the paragraph following the introductory paragraph. Verify that 8! = 40,320.

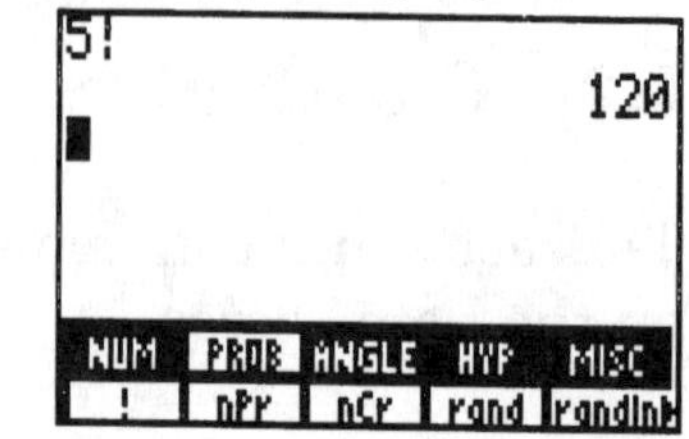

The permutation function $P(n,r)$ is the option shown as nPr ([F2]). This feature is used in the following manner: to compute $P(10,4)$, press 10 [F2] (nPr) 4 [ENTER] and the result will appear.

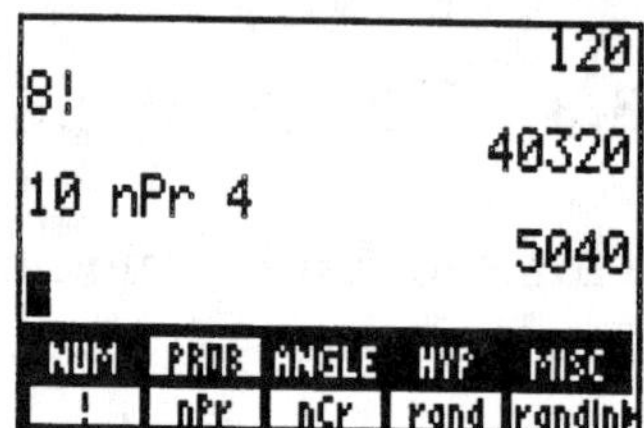

Combinations

The number of combinations of n items taken r at a time is given by $C(n,r) = \dfrac{n!}{r!\,(n-r)!}$. This feature on the TI-86 is given shown as nCr. Using Quick Example 2 in the text under Combinations of n items taken r at a time, you can compute $C(6,4)$ by pressing 6 [F3] (nCr) 4 [ENTER] to see 15 as the result.

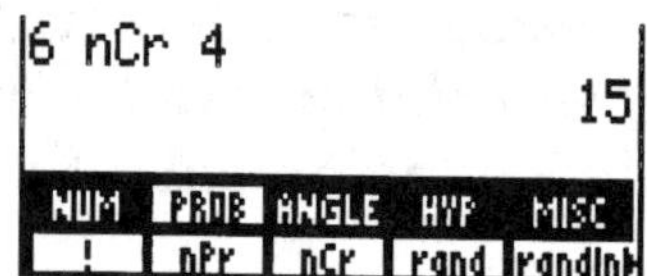

You're the Expert – Designing a Puzzle

There are no new calculator skills to speak of here.

CHAPTER 7

PROBABILITY

TI-86

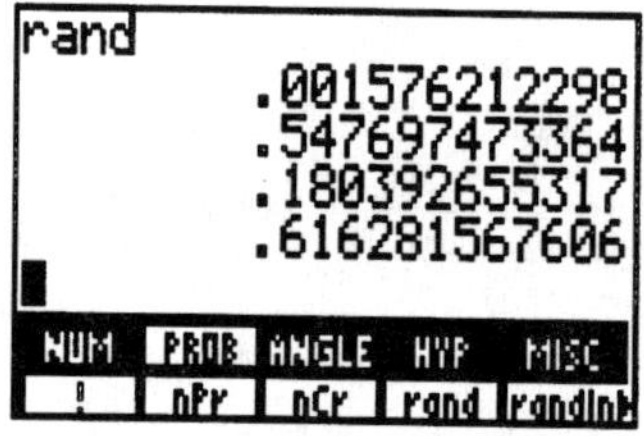

7.1 Sample Spaces and Events

This section requires no calculator skills.

7.2 Estimated Probability

Random number generators on the TI-86

The TI-86 is equipped with two random number generators, `rand` and `randInt`. The first (`rand`) produces a "random" decimal number between 0 and 1, and the latter (`randInt`) produces a "random" integer between two specified integer values.

Both of these commands are found under the math probability menu MATH PROB. Press [2nd] [MATH] [F2] (PROB) to see this menu.

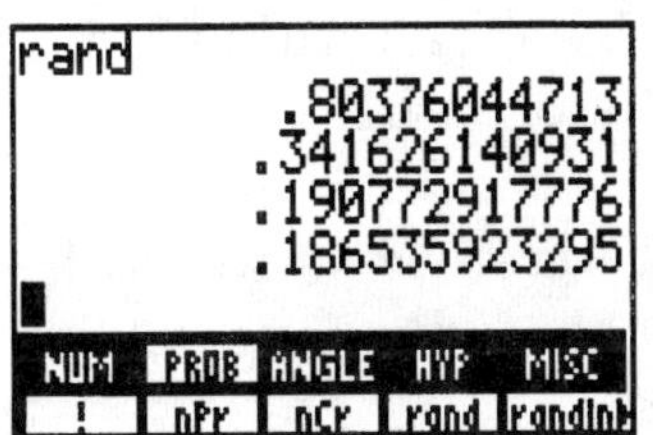

To see a random number between 0 and 1 generated, press [F4] (`rand`) [ENTER]. Press [ENTER] again and again to re-issue the command and generate more random numbers between 0 and 1. If you try this, the numbers generated by your calculator will probably be different than those shown on my screen at left.

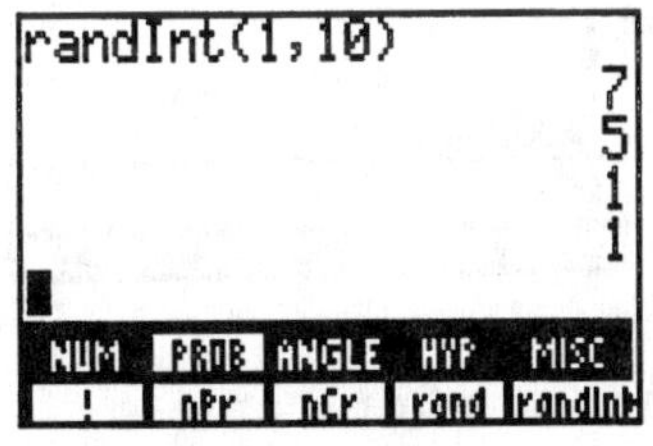

To use the random integer generator, say to generate a random integer between 1 and 10, with the MATH PROB menu accessed press [F5] (`randIn`) 1 [,] 10 [)] [ENTER]. Press [ENTER] again and again to re-issue the command and generate more random integers between 1 and 10.

In general, the command `randInt`(m,n) generates a random integer in the range $[m,n]$. The command `randInt`(m,n,k) generates k random integers in the range $[m,n]$. The k generated integers are returned in a list. For example, `randInt(2,8,5)` will generate 5 integers between the numbers 2 and 8.

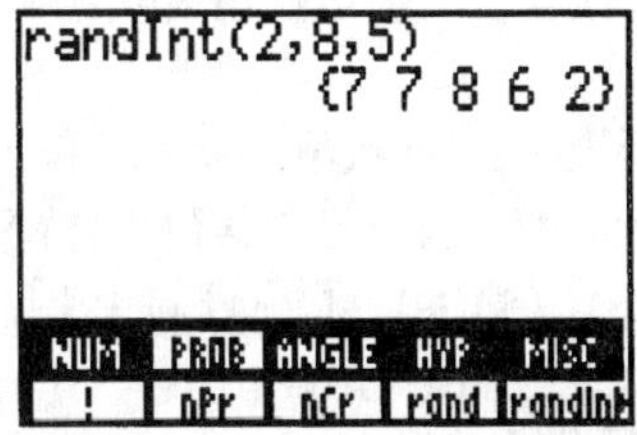

Example 3 *Simulating Experiments with Random Numbers*
In this example you are asked to use a simulated experiment to check the
following.
*(a) The estimated probability of heads coming up in a toss of a fair coin
approaches 1/2 as the number of trial gets large.*
*(b) The estimated probability of heads coming up in two consecutive tosses of a
fair coin approaches 1/4 as the number of trials gets large.*

For part (a), use 1 to represent heads and 0 to represent tails. You can use
the `randInt` command to generate a random integer between 0 and 1
(`rantInt(0,1)`).

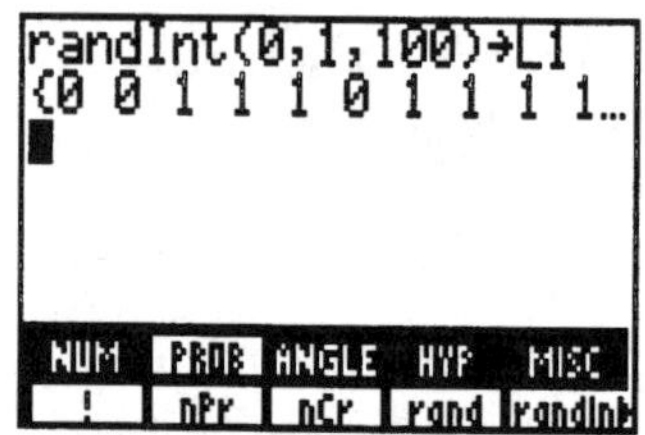

To do this a large number of times, say 100, and have
a record of the simulated outcomes, access the MATH
PROB menu (2nd [MATH] F2), and press F5 0, 1, 100)
STO▸ [L] ALPHA 1 ENTER. After a brief pause, 100
random integers between 0 and 1 (i.e., a total of 100
0's and 1's) will be produced and stored L1.

When the list is returned, you can use the right and left arrow keys to scroll
the list and view the outcomes on the home screen. But the advantage of
storing the outcomes in L1 is that when you view
them in the STAT editor (2nd [STAT] F2), you can keep
track of the outcome for a particular flip of the coin.
For example, the screen here shows L1(25)=1 at the
bottom left above the menu line, indicating that the
outcome of the 25th (simulated) flip was a head.

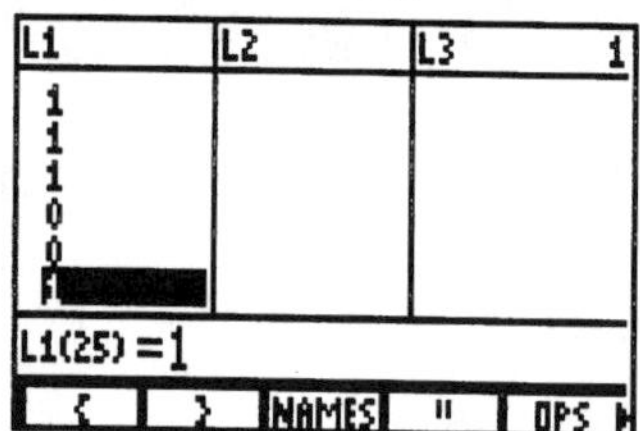

To complete the solution to part (a), you want to sum the list of outcomes
from the simulation. This amounts to finding the number of times a head
(the number 1) occurred.

The `sum` command is found in the miscellaneous
math menu MATH MISC. Press 2nd [MATH] F5 (MISC)
F1 (sum) ALPHA [L] 1 ENTER to compute the sum of list
L1. The outcomes of my simulation produced 51
heads, and thus the estimated probability of a heads

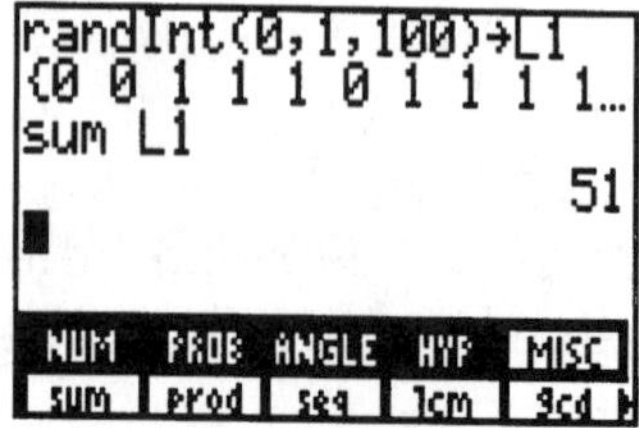

coming up is $\dfrac{51}{100} = 0.51$. Remember, the estimated probability that you obtain may be different.

You don't have to stop at 100 simulated flips of the coin. You can simulate flipping a coin for as many times as your calculator has memory to compute and store the result. For example, the screen shown at the left is the simulation of 1,000 flips of a coin, with the number of heads appearing. Thus for this simulation the estimated probability of obtaining a head is 0.476.

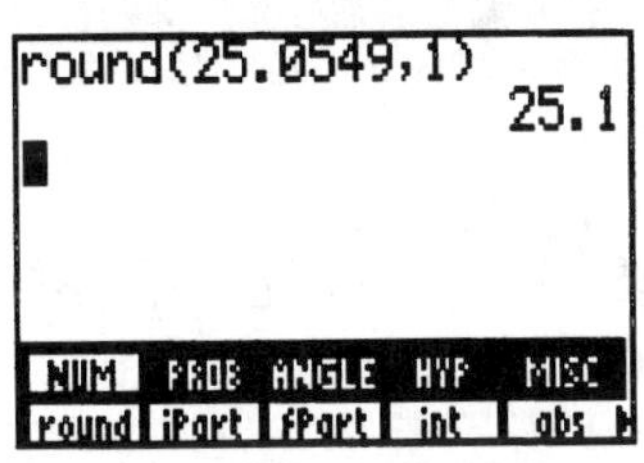

For part (b), you can use the command `round(rand,0)` to generate either a 0 or 1 by rounding the random number either down to 0 or up to 1. The command $round(X,n)$ will round the number X to n decimal places. The `round` command is found in the MATH NUM menu.

Adding two such commands will return a 0, 1, or 2 (can you explain why?), simulating the number of heads obtained in the flip of two coins. Dividing these numbers by 1/2 will give you 0, 0.5, or 1. The integer parts of these numbers (the part before the decimal) are 0, 0, and 1, respectively. Thus, the number 0 will indicate that something other than two heads occurred, and the number 1 will indicate that two heads did occur.

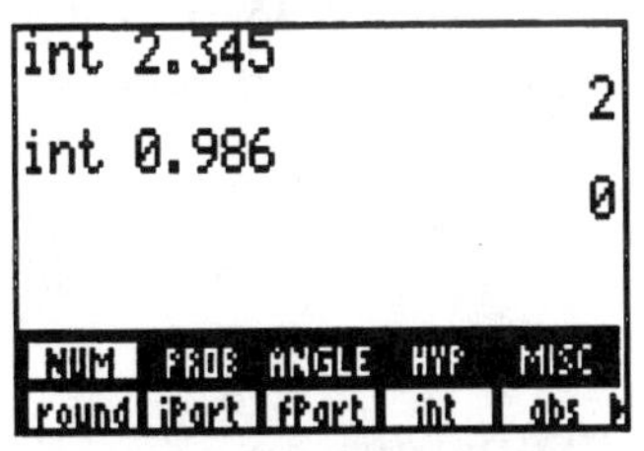

The calculator's `int` command (also found in the MATH NUM menu) will return the integer part of a number. For example, `int 2.345` will return the number 2, and `int 0.986` will return the number 0. Thus, `int` 0 = 0, `int` 0.5 = 0, and `int` 1 = 1.

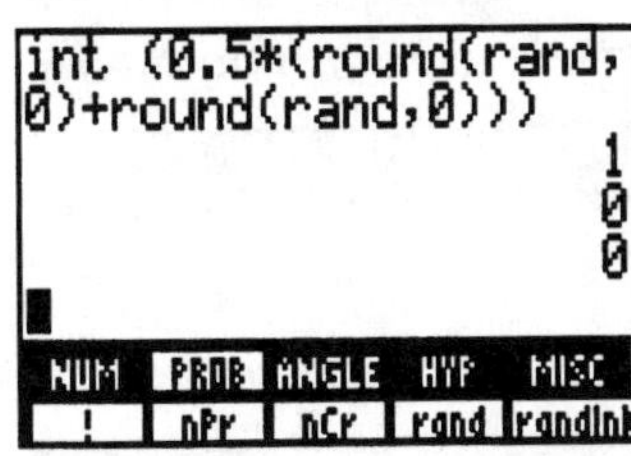

Therefore, if the command `int(0.5*(round(rand,0)+round(rand,0)))` returns a 1, this indicates that two heads *were* obtained, and a result of 0 will indicate that two heads were *not* obtained.

Access the equation editor ([GRAPH] [F1]) and define y1 to be this command. Press (carefully!) [2nd] [MATH] [F1] (NUM) [F4] (int) [(] 0.5 [×] [(] [F1] (round) [2nd] [MATH] [F2] (PROB) [F4] (rand) [,] 0 [)] [+] [2nd] [MATH] [F1] (NUM) [F1] (round) [2nd] [MATH] [F2] (PROB) [F4] (rand) [,] 0 [)] [)] [)]. Be sure to scroll and verify that the entry is defined just as shown in the above paragraph. When finished, exit to the home screen.

In the MATH MISC menu is a command to create a sequence of numbers called seq. To generate the sequence of numbers from 1 to 100, press [2nd] [MATH] [F5] (MISC) [F3] (seq) [x-VAR] [,] [x-VAR] [,] 1 [,] 100 [,] 1 [)] [STO▸] [X] [ENTER]. Note that you are storing the sequence to uppercase X (over [+]) and not to lowercase x [x-VAR].

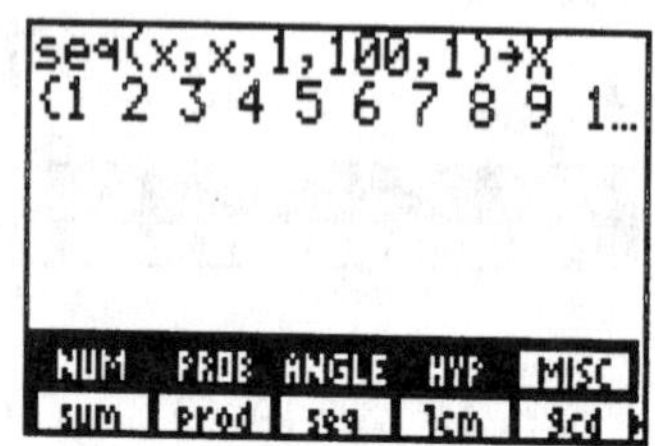

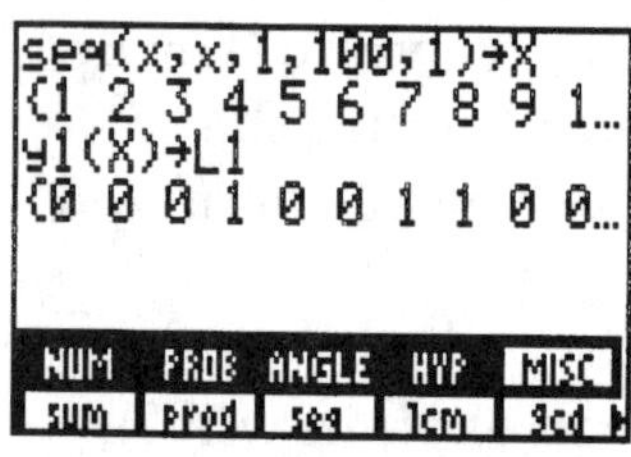

Now press [2nd] [alpha] [y] 1 [(] [ALPHA] [X] [)] [STO▸] [L] [ALPHA] 1. Press [ENTER] to obtain the list of outcomes of the simulation of two consecutive flips of a fair coin 100 times, and to place the outcomes in L1. Press [2nd] [STAT] [F2] (EDIT) to scroll and view the list of outcomes. How would you determine if two heads were observed on the 10th simulated trial of this experiment?

Finally, to complete the experiment, exit to the home screen. Press [2nd] [MATH] [F5] (MISC) [F1] (sum) [ALPHA] [L] 1 [ENTER] to count the number of times two heads appeared. Dividing by 100 indicates that the estimated probability of heads coming up in two consecutive tosses of a fair coin is about 0.35.

An Alternate Method – Using randInt

Another and possibly more convenient calculator method is using the randInt command. Recall that the randInt command is found in the MATH PROB menu.

On a clear home screen, input the command randInt(0,2,100)→L1 by pressing [2nd] [MATH] [F2] [F5] 0 [,] 2 [,] 100 [)] [STO▸] [L] [ALPHA] [1]. Press [ENTER]. This will randomly generate the numbers 0, 1, or 2 100 times and place (store) the results in the STAT editor as list L1.

The numbers 0, 1, and 2 indicate that no heads, one head, or two heads, respectively, were obtained on a particular simulated consecutive flip.

Now press [2nd] [STAT] [F2] (EDIT) to view the list of outcomes of the simulation. You now want to determine the number of times a 2 occurred.

To do this, use the right and up arrow keys to get to the very top of the

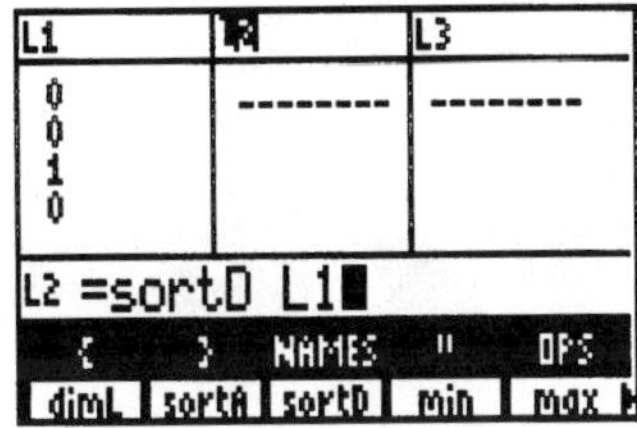

column titled L2. With the list name highlighted, press [F5] (OPS) [F3] (sortD) [ALPHA] [L] [1] so that L2=sortD L1 appears on the command line above the menu (see screen). This will sort list L1 in descending order (greatest value to least) and place the result in list L2. Press [ENTER] to see the sorted list.

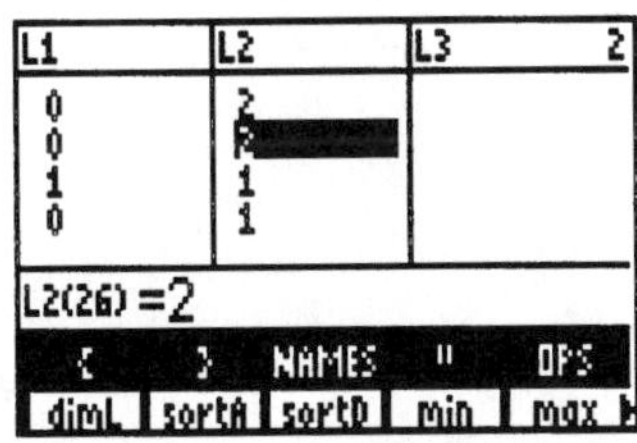

Use [↓] to scroll down L2 until you come to the first occurrence of the number 1. With the highlighter placed on the last number 2 occurring, you can see from the command line how many times two heads appeared. Thus, for my simulation you see that two heads appeared 26 times.

I can now conclude that the estimated probability of obtaining heads in two consecutive flips of a fair coin is $\dfrac{26}{100} = 0.26$. You will need to exit to the home screen to perform this calculation. Your results may (and should) be different.

The "random" Number Generator
The TI-86's "random" number generator is really a pseudo-random number generator in that you can control the sequence of numbers generated. This means that it is possible for you and I to obtain the very same numbers that are generated by the calculator.

To see how this is done, let's both type 25 [STO▸] [2nd] [MATH] [F2] (PROB) [F4] (rand) [ENTER]. This "seeds" the generator. With the MATH PROB menu displayed, press [F4] (rand) [ENTER] and we should get the same "random" number! Now press F5 (randInt) 0 , 2 , 5) ENTER and see that you get the exact same list of numbers as shown on the screen. Any number can be used as a seed, as long as we both use the same number.

7.3   Empirical Probability

There are no new calculator skills in this section. All computations can be
performed on the home screen.

7.4   Probability and Counting Techniques

There are no now calculator skills in this section. All values needed can be
computed using the $_nC_r$ command in the MATH PROB menu (see section 6.4 of
this manual), and all computations can then be performed on the home
screen.

7.5   Abstract Probability

There are no new calculator skills in this section.

7.6   Conditional Probability and Independence

There are no new calculator skills in this section.

7.7 Bayes' Theorem and Applications

There are no new calculator skills in this section. However, careful use of parentheses is needed to obtain correct answers. For instance, in Example 3 *Grades* from the text,

$$P(R_1 \mid A) = \frac{(0.1)(0.2)}{(0.1)(0.2) + (0.8)(0.6) + (0.9)(0.2)} \, .$$

See the screen to the right regarding how to compute this value on the home screen.

Note the set of parentheses added to begin and end the denominator. This indicates to the calculator that the sum of the products given in denominator should be treated as one calculation before it is divided into the product given in the numerator.

You're the Expert – The Monty Hall Problem

There are no new calculator skills needed for this section.

CHAPTER 8

RANDOM VARIABLES
AND
STATISTICS

TI-86

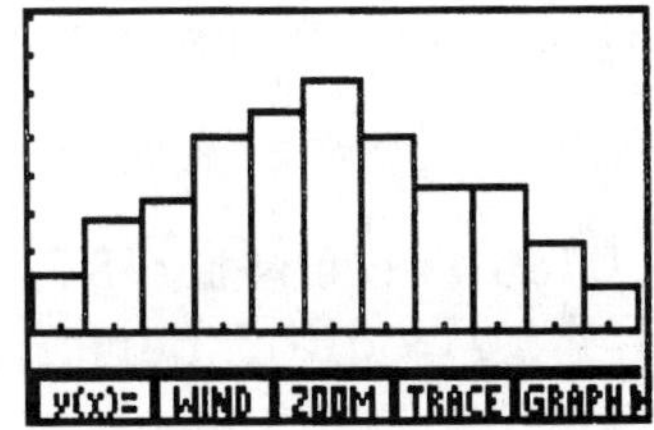

8.1 Random Variables and Statistics

In this section you will learn how to graph probability distributions as a histogram on the TI-86 when the distribution has already been determined. You will also see how to use a histogram to help develop a probability distribution.

Graphing Probability Distributions: Histograms

Consider the probability distribution developed in Example 3 *Empirical Probability Distribution* from the text. The distribution is reproduced below for convenience.

x	0	1	2	3
$P(X = x)$	$\dfrac{1}{8}$	$\dfrac{3}{8}$	$\dfrac{3}{8}$	$\dfrac{1}{8}$

To graph this distribution, press [2nd] [STAT] [F2] (EDIT) to enter the STAT editor. If you have data in list L1, arrow to the very top of the list, press [CLEAR] and then [ENTER]. This will clear the values from the list. Do the same for L2 if necessary.

Place the values of X (top row of the table) in L1 and the probabilities (second row of table) in L2. The numbers will convert to decimal form as you enter them. Press [GRAPH] [F2] (WIND) and set the window to $0 \le x \le 4$ with xScl=1 (width of the bars) and $0 \le y \le 0.5$ with yScl=0.1.

Then press [2nd] [STAT] [F3] (PLOT) to access the STAT PLOT editor. Now that you have accessed the STAT PLOT editor, perform the following procedure.

1. Press [F1] to select PLOT1.
2. Press [ENTER] to turn the plot on.
3. Press [▼] and then [F4] (HIST) to select histogram as the plot of choice.
4. Press [▼] and set Xlist Name equal to L1. You may have to type [L] [ALPHA] [1] to do this, or you

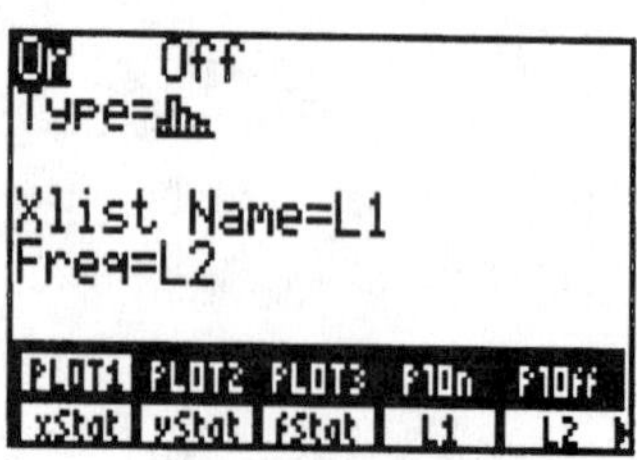

can select L1 from by pressing the appropriate F-key in the name menu
(at the bottom of the screen) where L1 is located.

5. Press $\boxed{\blacktriangledown}$ and set Freq equal to L2.
6. Press $\boxed{\text{GRAPH}}$ $\boxed{\text{F5}}$ (GRAPH). You should obtain the
   graph of the distribution, represented as a
   histogram.

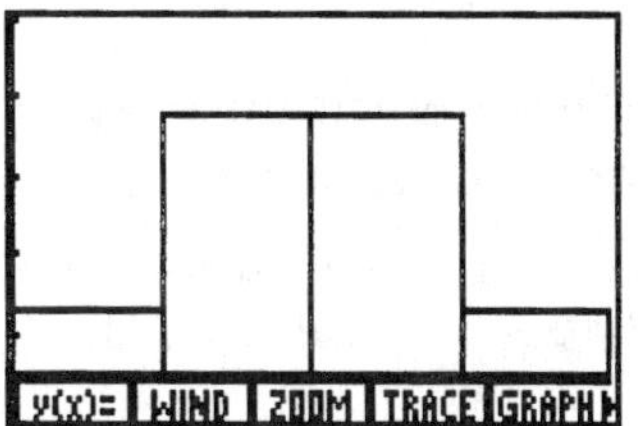

Note that the menu line is cutting off the view of the bottom of your
histogram. To see the entire graph, press $\boxed{\text{CLEAR}}$ and the menu line will be
removed from the graph screen. To bring the menu line back, press $\boxed{\text{GRAPH}}$.

Another way to avoid this problem is to set yMin to a value less than 0. An
appropriate minimum value for y would be about
negative one-fourth of the maximum value used for y.
Thus, since yMax=0.5, you can set yMin to be –0.5/4
= –0.125. At the yMin= prompt just type $\boxed{\text{(-)}}$ 0.5 $\boxed{\div}$ 4
$\boxed{\text{ENTER}}$ and the value will automatically be simplified.
Now press $\boxed{\text{GRAPH}}$ and see the entire graph along with
the menu line.

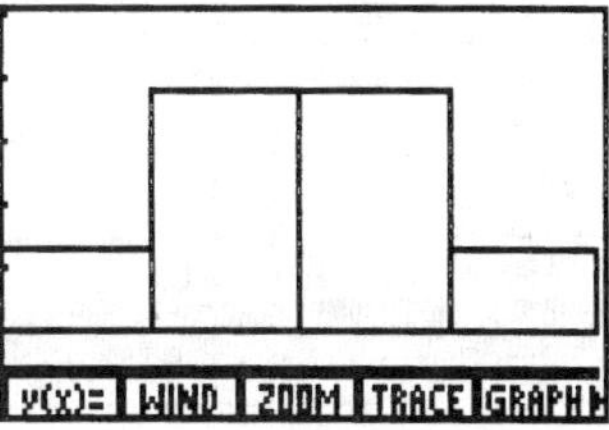

Developing a Probability Distribution from a Histogram

Suppose you want to develop a histogram on your TI-86 without first
computing the probability distribution. You can not only do this, but you can
then use the histogram to develop the distribution!

Recall from chapter 7 the calculator's *sequence* command seq. This command
allows you to define a sequence of numbers in any way you wish. This
command is found in many locations, including the MATH MISC menu ($\boxed{\text{2nd}}$
$\boxed{\text{MATH}}$ $\boxed{\text{F5}}$). The syntax for the command is
seq(expression,variable,begin,end,step).

For example, if you want to generate the sequence of
values for the expression $2x - 5$ as the variable x
ranges from integers –1 through 6, you can input the
command seq(2x-5,x,-1,6,1) and press $\boxed{\text{ENTER}}$ to
view the result. Note that $2x - 5$ is the expression, x
is the variable, we began with –1, ended with 6, and

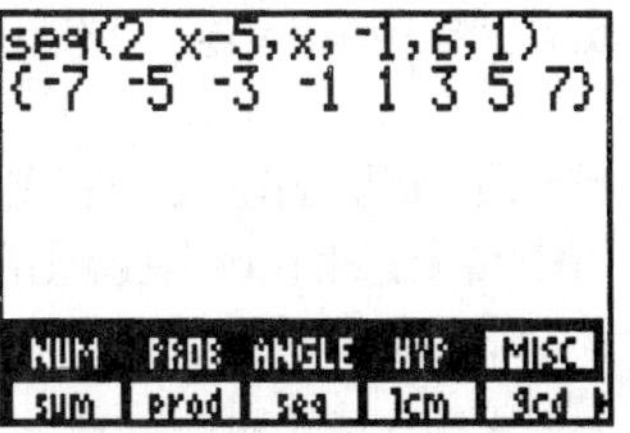

went from −1 to 6 in steps of 1.

If you wanted to go from −1 to 6 in steps of 2, you need to input `seq(2x-5,x,-1,6,2)` and press ENTER. This uses only the values −1, 1, 3, and 5 for the variable x in the expression.

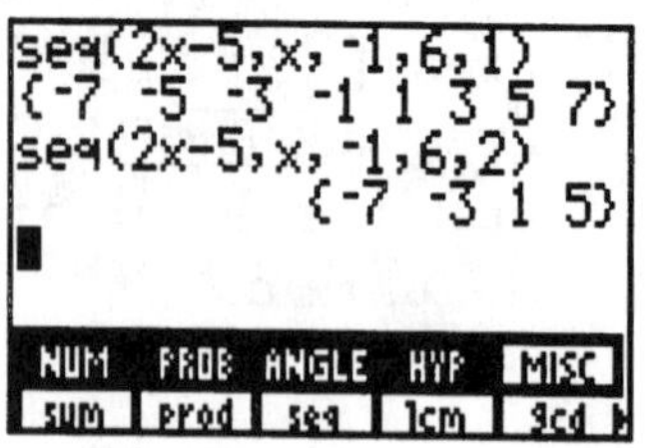

Consider Example 5 *Nutrition* from the text. Place the 30 individual percentages in L1. Since each individual percentage in L1 is 1/30th of the total number of percentages listed, you want to place 1/30 in list L2 to correspond with each value in L1.

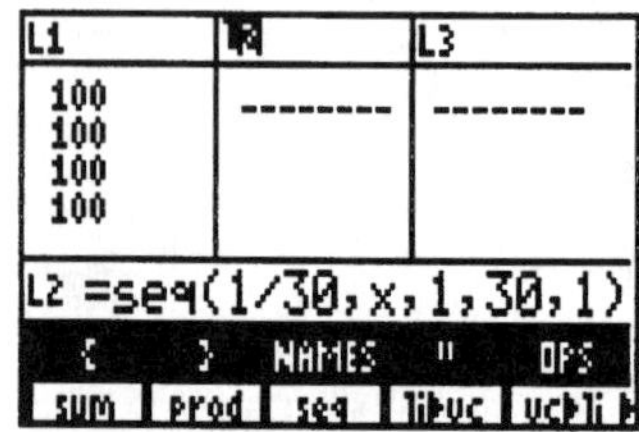

Do this by using your arrow keys to get to the top of list L2, and while there press F5 (OPS) MORE F3 (seq) 1 ÷ 30 , x-VAR , 1 , 30 , 1). The command line should look like what is shown on the screen to the left. After pressing ENTER, list L2 should look like the screen shown at the right.

Now press GRAPH F2 (WIND) and set the window to 25 ≤ x ≤ 105 with xScl=10 (width of the bars) and −0.5/4 ≤ y ≤ 0.5 with yScl=0.1. Afterwards, access the STAT PLOT editor (2nd [STAT] F3 (PLOT)) and go through the six steps given in Example 3 *Empirical Probabilities* described earlier in this section to obtain a histogram of the distribution.

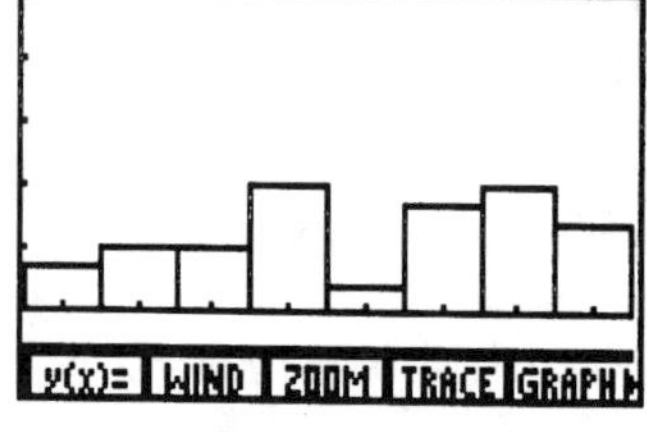

Note that since yScl=0.1, we can see (by counting tick marks) that no probability is greater than 0.3, so we can adjust our graph window to −0.3/4 ≤ y ≤ 0.3 for a better view of the histogram.

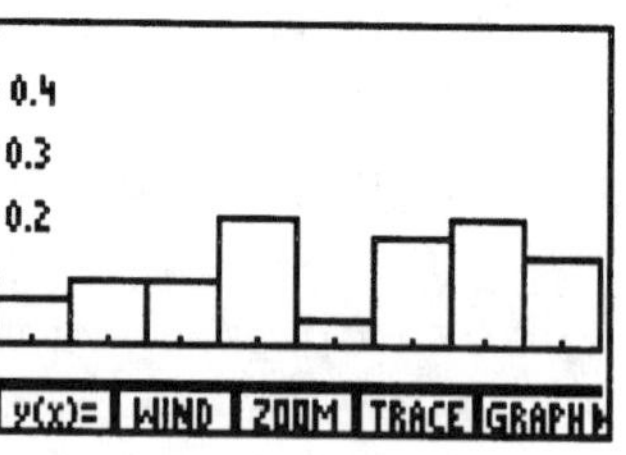

Each tick-mark on the x-axis in the center of the rectangles represents the value in the center of the interval. For example, the first rectangle is from 25 to 35 (since xMin=10 and xScl=10), so the first tick-mark represents 30. This means that 30 is a possible value of X in the probability distribution.

212

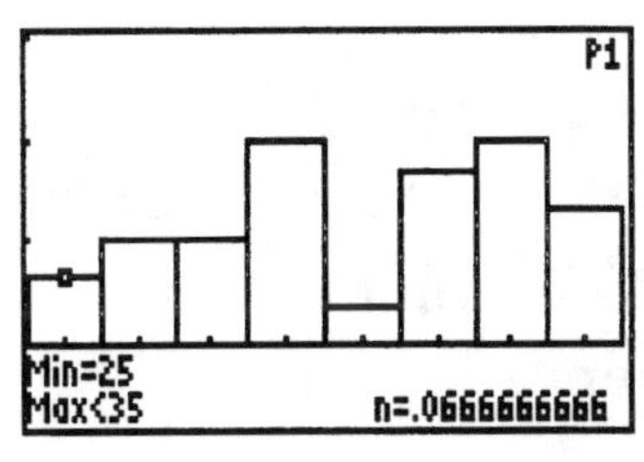

Press [F4] (TRACE). A cursor appears at the top center of the first rectangle. Also, the left and right endpoints of the rectangle are given (Min and Max) along with a value for n. The midpoint of the interval is a possible value of X, and the value of n is the corresponding probability for this outcome, i.e., $P(X = 30) \approx .0667$!

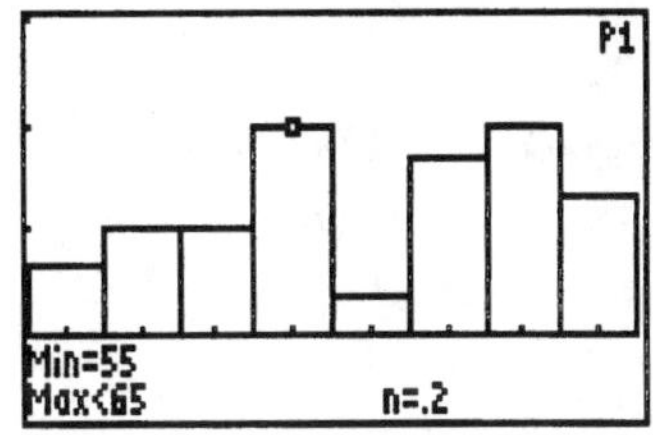

You can compete the probability distribution by using the right arrow to trace the histogram and record the outcomes X together with their corresponding probabilities n. For example, the screen at the left indicates that $P(X = 60) = 0.2$.

The exact probabilities associated with the possible outcomes for X can be found using a similar (and possibly simpler) procedure to develop a frequency distribution instead of a probability distribution. This is done by simply changing the PLOT1 Freq option from L2 to 1, i.e., Freq=1 (press [ALPHA] 1 at the Freq prompt). You would also have to change the y-values in your window setting to $-8/2 \le y \le 8$ with yScl=1.

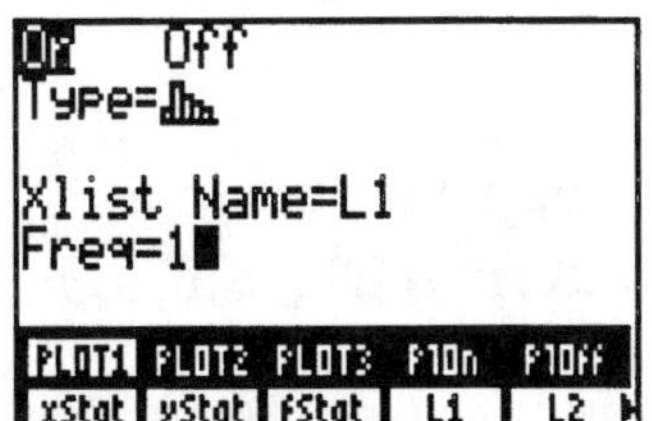

Graphing the plot will produce the histogram shown in the screen. Note that this histogram is the same as the one used to develop the probability distribution of X. Using TRACE and [▶] you will see that the values of n are now integers, representing the frequency of the occurrence of the value of X.

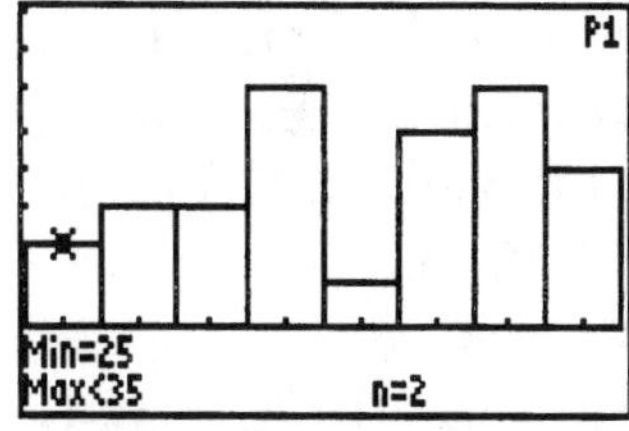

The screen graph shown above indicates that $X = 30$ appears 2 times, and since there are 30 total values you can conclude that $P(X = 30) = \dfrac{2}{30} = 0.06667$. Use your right arrow key to trace the histogram and complete the frequency distribution and hence the probability distribution with exact probabilities.

8.2 Bernoulli Trials and Binomial Random Variables

As stated in the text, if X is the number of successes in a sequence of n independent Bernoulli trials, then $P(X = x) = C(n,x)\, p^x q^{n-x}$, or for the TI-86 this would be $P(X = x) = nCr\, p^x q^{n-x}$. Recall (from chapter 6) that the combination function nCx is found in the `MATH PROB` menu.

Example 2 *Will You Need Me When I'm 64?*
This example from the text states the following: *The probability that a randomly chosen person in the US is 65 or older is approximately 0.2.*
(a) *What is the probability that, in a randomly selected sample of 6 people, exactly 4 of them are 65 or older?*
(b) *If X is the number of people of age 65 or older in a sample of 6, construct the probability distribution of X and plot its histogram.*
(c) *Compute $P(X \le 2)$.*
(d) *Compute $P(X \ge 2)$.*

Part (a) can be solved using the TI-86 simply by recognizing that this is a direct application of the binomial formula $P(X = x) = C(n,x)p^x q^{n-x}$, where $n = 6$, $x = 4$, $p = 0.2$, and $q = 1 - 0.2 = 0.8$.

On the home screen, press 2nd [MATH] F2 (PROB) to access the combination function in the math probability menu. Then press 6 F3 (nCr) 4 × 0.2 ∧ 4 × 0.8 ∧ 2 ENTER to obtain the result. Thus $P(X = 4) = 0.01536$.

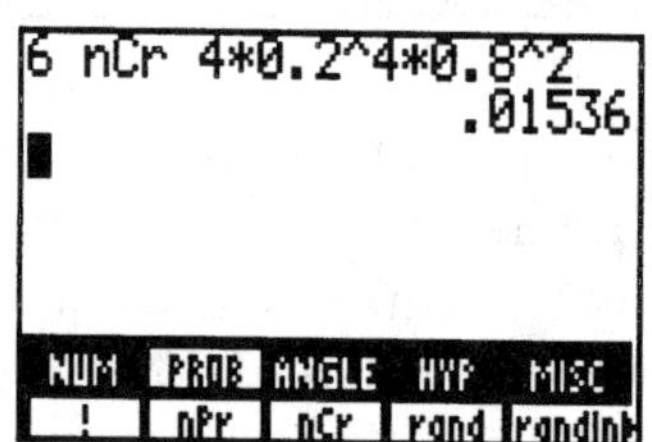

For part (b), the distribution can be developed in multiple ways using the TI-86. One way is to use the calculator's `TABLE` feature. The other is to use the `STAT` editor.

Using TABLE
Enter the binomial distribution function into the equation editor (GRAPH F1) as `\y1=6 nCr x*0.2^x*0.8^(6-x)`. Before leaving the equation editor, you may have to turn off any plots that are on. If any of the plots are highlighted at the top of the

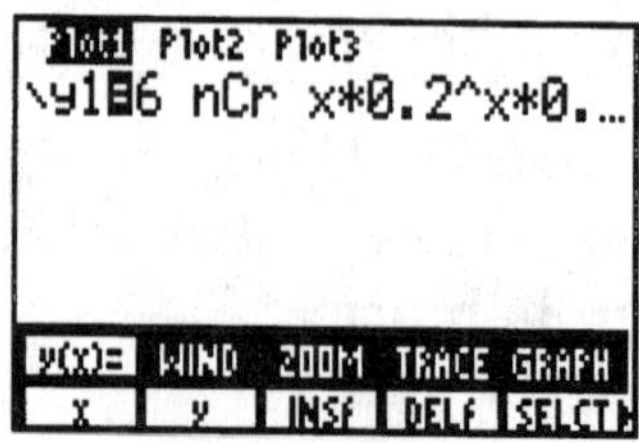

equation editor screen, arrow up there, place the cursor over that particular plot and press [ENTER]. You can press [▼] to return to equation editing.

Now press [TABLE] [F2] (TBLST) and set `Indpnt:` to `Ask` (the values for `TblStart` and `ΔTbl` don't matter for this option). Press [F1] (TABLE) and enter the values X = 0, 1, ..., 6, pressing [ENTER] after each number. You are only allowed to enter six values for x, so you will have to type 6 over the 5 to get the last probability.

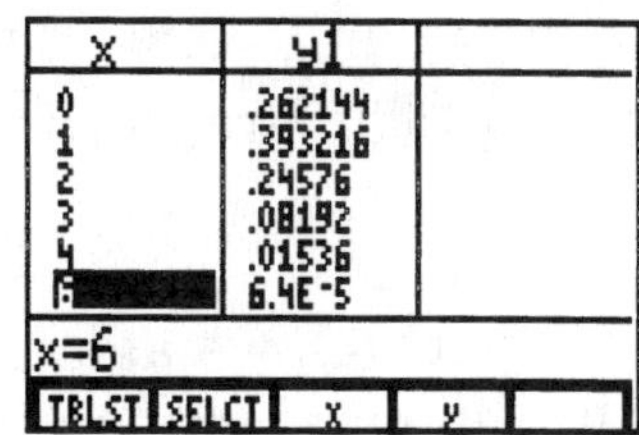

Using the STAT Editor

This method is nice because it will help you carry out both directives mentioned in the problem, namely construct the distribution of X and plot its histogram.

Make sure that the binomial function has been assigned to y1 in the equation editor in the manner shown above in the TABLE method. Go into the STAT editor ([2nd] [STAT] [F2]) and clear out lists L1 and L2 (arrow to the top of the list and press [CLEAR] [ENTER]). Enter the values of X (0, 1, ...,6) in L1.

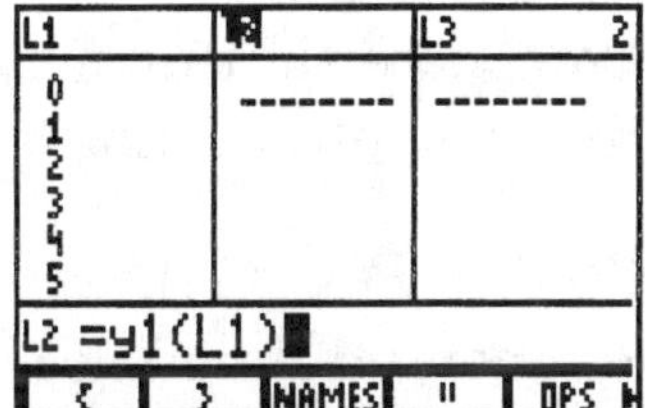

Arrow to the top of list L2 and press [2nd] [alpha] [y] 1 [(] [ALPHA] [L] 1 [)]. This defines list L2 as the values of the binomial function defined in y1 for each x given in L1 (see section 1.1 *Functions from the Numerical and Algebraic Viewpoints* for evaluating functions using the TI-86). Press [ENTER] and the distribution is constructed. It is now just a matter of recording the results in a table on your paper.

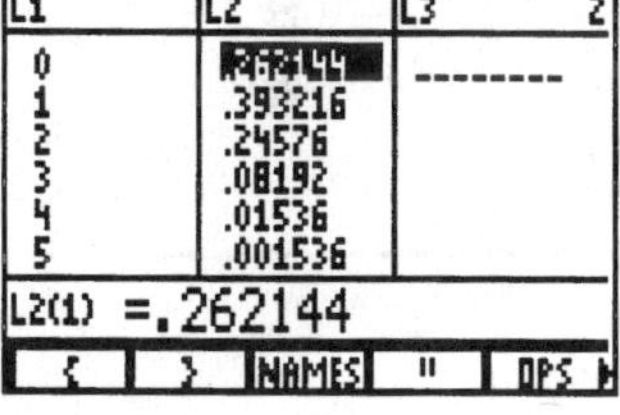

Now that the distribution is in the STAT editor, you have to define the plot options and set the graph viewing window. Press [2nd] [STAT] [F3] (PLOT) to access to STAT PLOT editor. Follow the procedure given in section 8.1 under *Graphing probability distributions: Histograms*. Your screen should look the screen shown here.

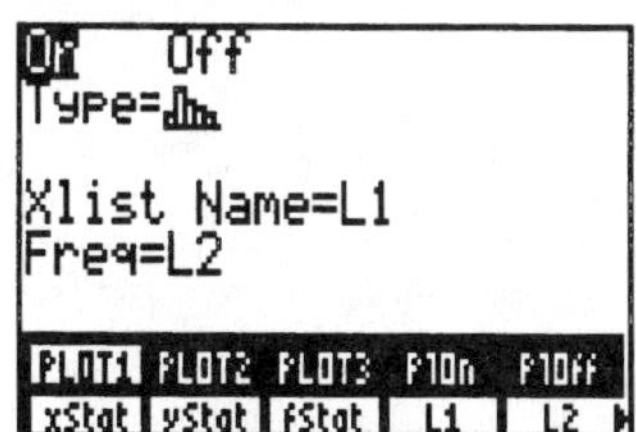

After setting up the PLOT editor, press [GRAPH] [F2] (WIND) and define the graph window as $0 \le x \le 7$ with xScl=1 and $-0.5/4 \le y \le 0.5$ with yScl=0.1. Finally, press [F5] to graph the distribution.

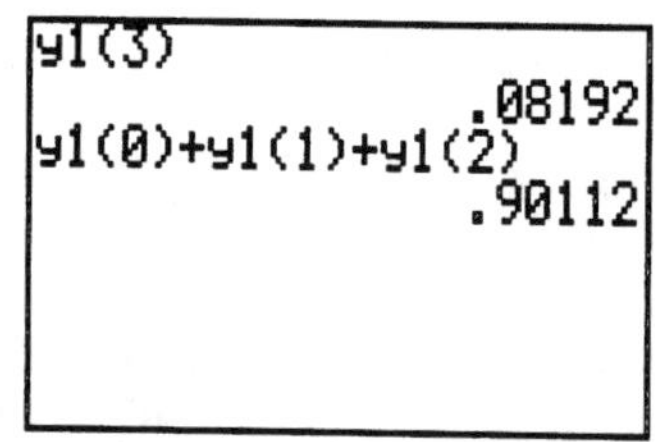

For part (c), we again use the convenience of having the binomial probability function defined in the equation editor as y1. If we want to compute the probability that $X = 3$, we can just evaluate the function y1 at $x = 3$ on the home screen by pressing [2nd] [alpha] [y] 1 [(] 3 [)] [ENTER].

Knowing that $P(X \le 2) = P(X = 0) + P(X = 1) + P(X = 2)$, on the home screen we can enter this as y1(0)+y1(1)+y1(2) and press [ENTER] for the result. As above, you need lowercase y to reference the function. Thus, $P(X \le 2) = 0.90112$.

Part (d) is similar to part (c), as you can enter on the home screen the expression y1(2)+y1(3)+y1(4)+y1(5)+y1(6). However, note that using the *complement* would require fewer computations. Then we can write
$$P(X \ge 2) = 1 - P(X < 2) = 1 - (P(X = 0) + P(X = 1)).$$
Thus, only two calculations of the binomial function is needed verses five. On the home screen enter 1-(y(0)+y(1)) and press [ENTER] to see the result.

8.3 Measures of Central Tendency

The mean of a number can easily be computed on the home screen, especially for relatively small data sets. The median value of a data set requires the values to be arranged in ascending (increasing) order. Let's see how this can be done using the data values given in Example 1 *Teen-age Spending* from the text.

Access the STAT editor (2nd [STAT] F2) and clear any lists having data in them. Place the data values in list L1. When finished, arrow to the top of list L1, press F5 (OPS) and then F2 (sortA) to select sort ascending. Then press ALPHA [L] 1 ENTER to perform the sort. You can now scroll down the list and identify the middle value(s).

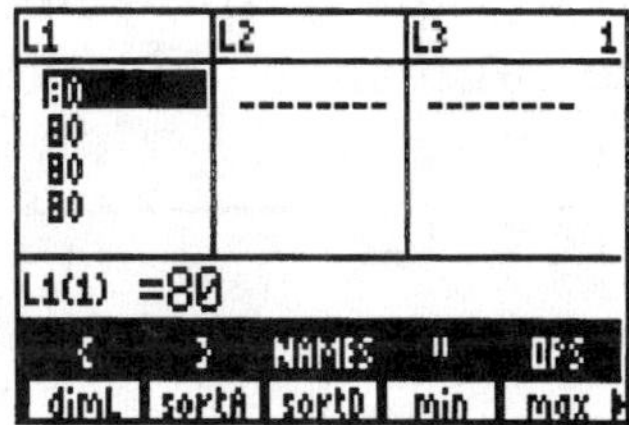

Expected Value of a Random Variable

The expected value $E(X)$ of a random variable X can again be computed on the home screen for a relatively small number of possible outcomes for X. A convenient way of computing the expected value of a random variable X in general is by using lists and performing the necessary arithmetic on the lists.

Let's use Example 3 *Sports Injuries* from the text to illustrate this. In the STAT editor, clear lists L1 and L2. Place the values of X (number of injuries) in L1 and the corresponding probabilities in L2. Press EXIT to return to the home screen.

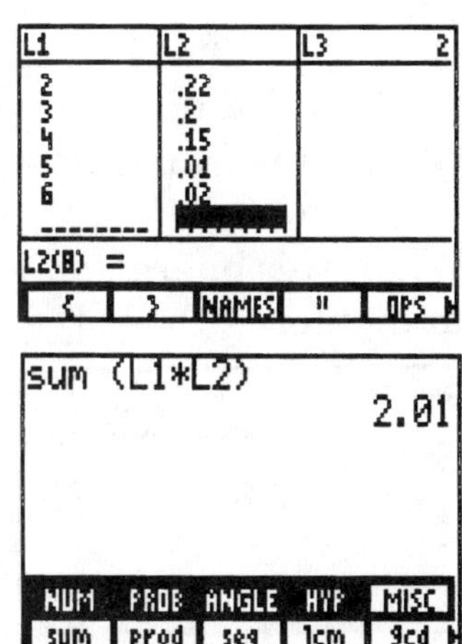

Since the expected value is $E(X) = \sum x \cdot P(X = x)$, on the home screen you will sum the product of L1 and L2 by pressing 2nd [MATH] F5 (MISC) F1 (sum) ((ALPHA [L] 1 × ALPHA [L] 2)) ENTER. The result indicates that $E(X) = 2.01$ injuries.

8.4 Measures of Dispersion

The standard deviation and variance of a data set are important values associated with the set, but rarely is it convenient to compute one (or the other) of these values on the home screen directly from the formula without the aid of lists. You will learn in this section how to use the TI-86 to find the standard deviation of a set of values as well as for a random variable.

The values 28, 29, 30, 31, 31, 33, 30, and 28 are found in Example 1 *Television Ratings* in the text. Let's use these values to find the standard deviation and variance.

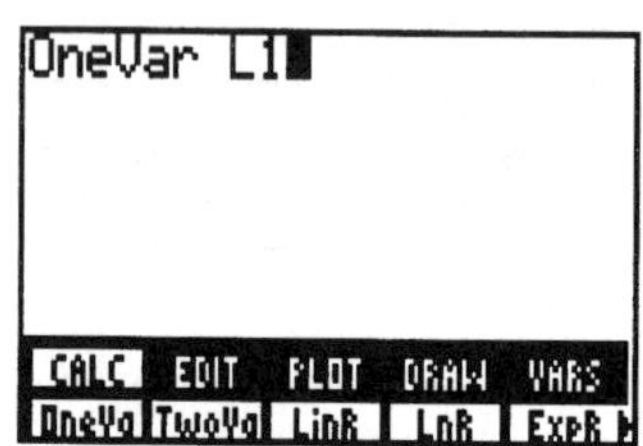

Enter the STAT editor and clear lists L1 and L2, if necessary. Input the data values into list L1. Once the values are inputted into the list, press [EXIT] to return the home screen. Now press [2nd] [STAT] [F1] (CALC) [F1] (Oneva) [ALPHA] [L] 1 [ENTER] to obtain a list of summary statistics for this data set.

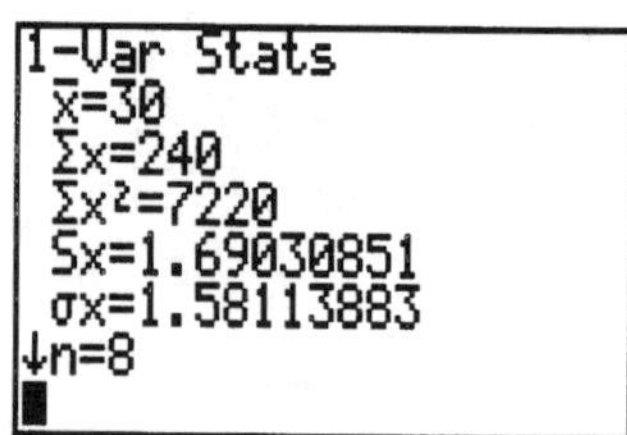

The summary statistics shown include the mean $\bar{x}$, the sample standard deviation Sx, the population standard deviation σx, the number of values in the data set n, and the median Med. The arrows ↓ and ↑ appearing on the screen indicate that there is more to the list. The rest of the list can be viewed by using your up and down arrows. You can press [EXIT] as often as needed to remove the menus from the screen and view more of the list.

When this operation is performed, all of these values are automatically stored in memory. Press [2nd] [STAT] [F5] (VARS) to access the STAT variables menu.

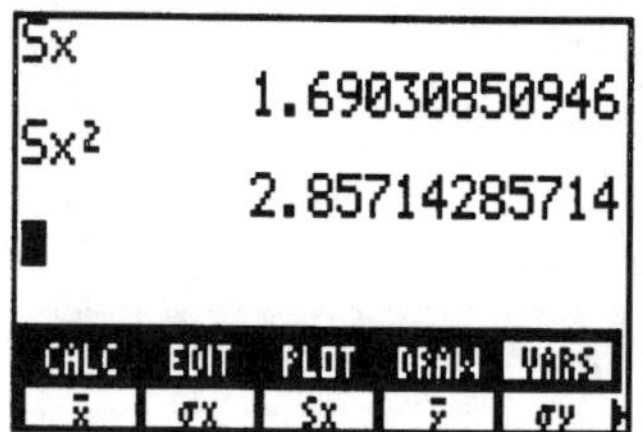

There you will see the statistical variables that were most recently computed. In particular, [F3] will access the sample standard deviation Sx. Press [CLEAR] (to clear the home screen) and then [F3] [ENTER] to see this value. You can obtain the sample variance by pressing [F3] (Sx) [x^2] [ENTER].

218

Standard Deviation of Random Variable

Let's turn our attention to finding the variance and standard deviation of a random variable. We will again consider an example from the text, and we will illustrate two methods for finding the desired quantities.

Example 3 *Variance of a Random Variable*
Compute the variance and standard deviation for the following probability distribution.

x	10	20	30	40	50	60
$P(X = x)$	0.2	0.2	0.3	0.1	0.1	0.1

Method 1 – Using the Formula and Lists
Once more we will use the spreadsheet capabilities of the TI-86 to calculate the requested values. Clear all data from lists L1 and L2. Enter the values for X into L1 and the corresponding probabilities into L2. Press [EXIT] to return to the home screen.

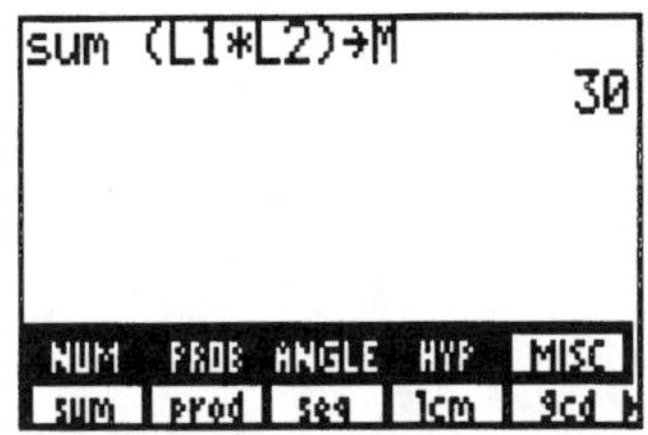

To find the mean μ of X press [2nd] [MATH] [F5] (MISC) [F1] (sum) [(] [ALPHA] [L] 1 [×] [ALPHA] [L] 2 [)] [STO▸] [M] [ENTER]. This command not only computes the mean μ of the random variable X but also stores the value as M for easy access.

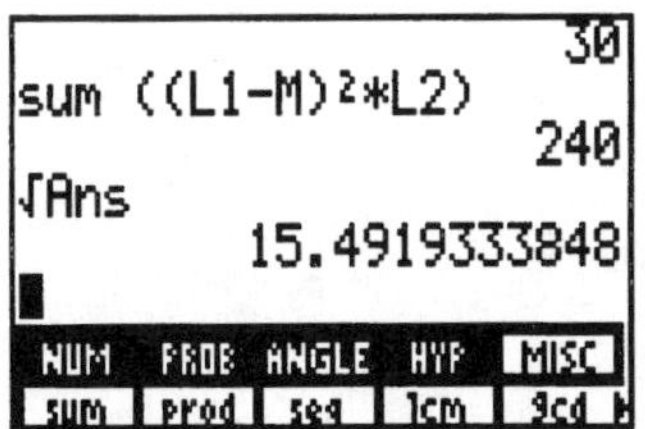

Now to find the variance, enter the expression sum ((L1-M)²*L2) and press [ENTER]. Finally, press [2nd] [√] [2nd] [ANS] [ENTER] to compute the standard deviation. Thus the variance $\sigma^2 = 240$ and the standard deviation $\sigma = 15.49$.

Method 2 – Using OneVar from the STAT CALC Menu
With X defined as L1 and $P(X = x)$ as L2, on the home screen press [CLEAR] (to clear the home screen) and then press [2nd] [STAT] [F1] (CALC) [F1] (OneVa) to paste OneVar on the home screen. Now enter L1 , L2 so that the statement OneVar L1 , L2 appears on the screen.

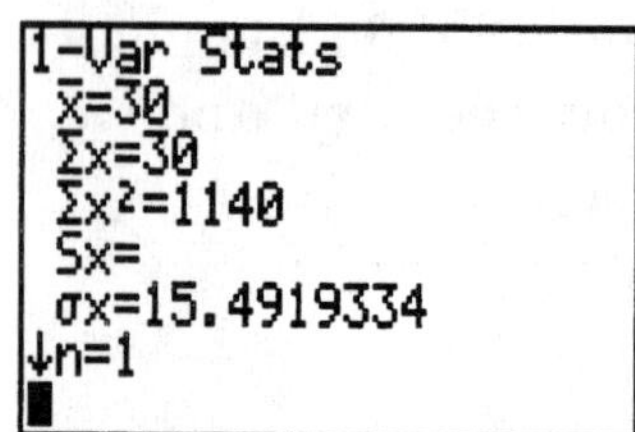

Press $\boxed{\text{ENTER}}$ to obtain the result. Press $\boxed{\text{EXIT}}$ twice to remove the menus. You now see that the mean $\mu = 30$ and that the standard deviation $\sigma = 15.49$.

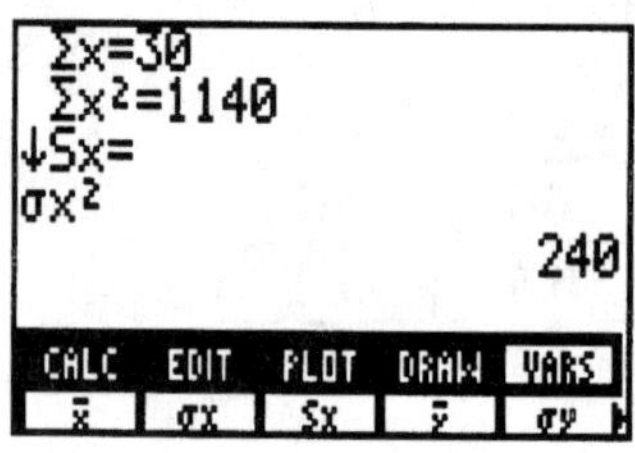

To find the variance of the random variable X, press $\boxed{\text{2nd}}$ [STAT] $\boxed{\text{F5}}$ (VARS) $\boxed{\text{F2}}$ (σX) $\boxed{x^2}$ $\boxed{\text{ENTER}}$.

8.5　Normal Distributions

The TI-86 does not have the built-in features needed to perform the computations necessary for this section. However, you can download a Statistics package from the Texas Instruments web site (www.ti.com). This package, like the Finance package, will give the TI-86 the ability to function much like the TI-83 and therefore make it more practical to perform the calculations needed for this section.

You're the Expert – Spotting Tax Fraud with Benford's Law

There are no new calculator skills or techniques in this section.

CHAPTER 9

MARKOV SYSTEMS

TI-86

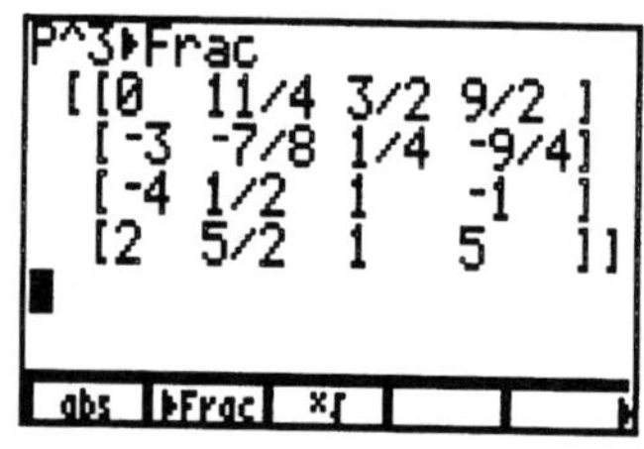

9.1 Markov Systems

There are no new calculator skills for this section.

9.2 Distribution Vectors and Powers of the Transition Matrix

There are really no new calculator skills for this section. Matrix multiplication is the skill needed here, and the TI-86 ability to perform this operation is discussed in section 3.2 of this manual. It would be a good idea for you to review this section if needed. We will, however, show you how you can perform repeated multiplication using the calculator's last answer feature Ans and hence the calculator's recursion feature.

Using Example 1 *Laundry Detergent Switching* from the text, with the initial distribution vector $v = [800 \quad 600]$ and transition matrix $P = \begin{bmatrix} 0.8 & 0.2 \\ 0.1 & 0.9 \end{bmatrix}$, you can find the distribution after several steps rather simply. Input these matrices into the TI-86 as matrices V and P, respectively ([2nd] [MATRX] [F2] (EDIT) gets you into the matrix editor). Remember to verify on the home screen that the matrices have been defined correctly.

Find the distribution after 1 step, the vector $v \cdot P$, by pressing [ALPHA] [V] [×] [ALPHA] [P] [ENTER]. To find the distribution after 2 steps, the vector $v \cdot P^2$, press [×] [ALPHA] [P] [ENTER]. This has the effect of multiplying the previous answer (the calculator's last answer, the distribution vector after 1 step) by the transition matrix P. In other words, you have computed the product $(v \cdot P) \cdot P = v \cdot P^2$.

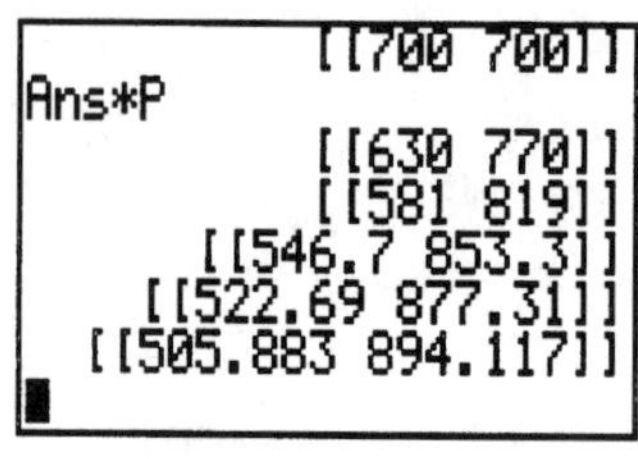

Press [ENTER] and the calculator will perform the same operation again, i.e., multiply the last (most recent) answer by P. This will give you the distribution after 3 steps, namely the vector $(v \cdot P^2) \cdot P = v \cdot P^3 = [581 \quad 819]$. Continuing to press [ENTER] will give you the distributions after any number of steps.

In Example 2 *Investments*, with transition matrix $P = \begin{bmatrix} 0.9 & 0.1 & 0 \\ 0.1 & 0.7 & 0.2 \\ 0 & 0 & 1 \end{bmatrix}$, you were asked to do the following three things.

(a) Write the two- and three-step transition matrices.
(b) Find the probability that The Solid Trust Saving and Loan, which is presently operating at a loss, will be bankrupt in each of the next three years.
(c) If half of the S&L's in the country are profitable and half are losing money, determine how many of them we can expect to be bankrupt after three years.

To complete part (a), input the one-step transition matrix P into your calculator and verify on the home screen that its entries were inputted correctly.

Since the two-step transition matrix is $P \cdot P = P^2$, press [ALPHA] [P] [x²] [ENTER] to obtain this matrix. Then press [ALPHA] [P] [^] 3 [ENTER] to find the three-step transition matrix $P^2 \cdot P = P^3$. Be sure to write these matrices on your paper as you find them. Note the ease at which you can find powers of a matrix.

To find the probability that *any* S&L that is currently operating at a loss will be bankrupt in each of the next three years (part (b)), you must look at the 2,3 entry of the one-, two-, and three-step transition matrices P, P^2, and P^3, respectively (why?). These entries are, in order, 0.2, 0.34, and 0.44 (see the screen in the previous paragraph). Thus, The Solid Trust S&L has a 20% chance of being bankrupt after one year, a 34% chance after two years, and a 44% chance after three years.

For part (c), the initial distribution vector is given by $v = [0.5 \quad 0.5 \quad 0]$, and as mentioned in the text you must find the product $v \cdot P^3$. Input v into the TI-86, naming it V.

After verifying on the home screen that V has been defined correctly, enter [ALPHA] [V] [×] [ALPHA] [P] [^] 3 and press [ENTER]. Look at the third entry of the resultant matrix to obtain the solution. Hence, we can expect 24.6% of the S&L's to be bankrupt after three years.

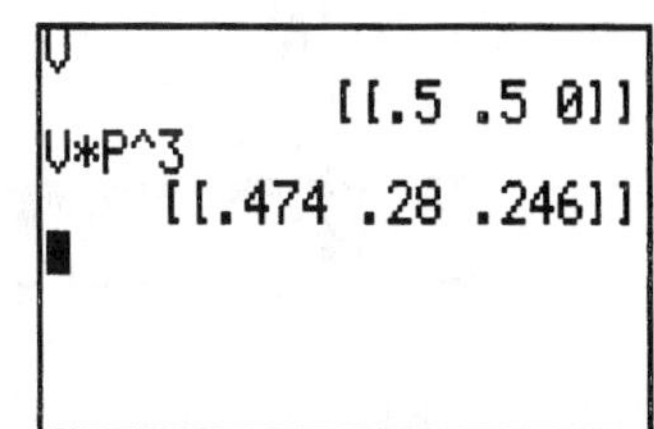

Using Technology
Under the *Using Technology* segment of this section in the text, you are encouraged to use technology (the TI-86, perhaps?) to calculate higher and higher powers of P and see if you notice anything interesting happen. To calculate higher powers of P means to find P^{10}, P^{25}, P^{50}, P^{100}, and so on.

Start doing this by finding P^5. Note that the entries of the resultant matrix are too long for all of them to fit on the screen. If you wanted to compute these entries to three decimal places, you could change the decimal mode setting to 3 ([2nd] MODE [▾] [▸] [▸] [▸] [▸] [ENTER]). But this will represent every entry to three decimal places; even the entry "2" would be represented as 2.000.

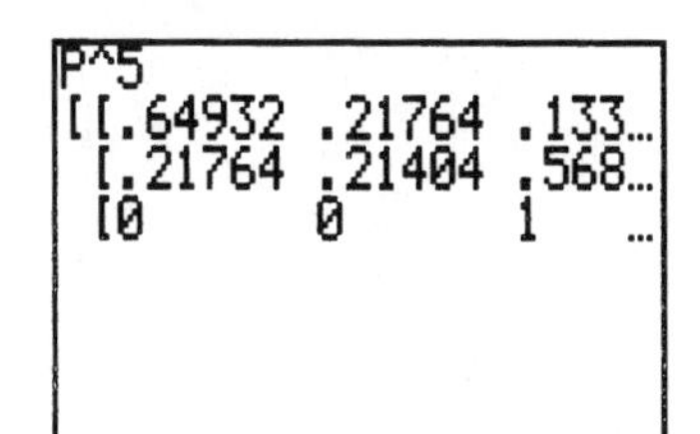

Another way that would produce possibly more aesthetically appealing results would be to use the rounding feature in the MATH NUM menu ([2nd]

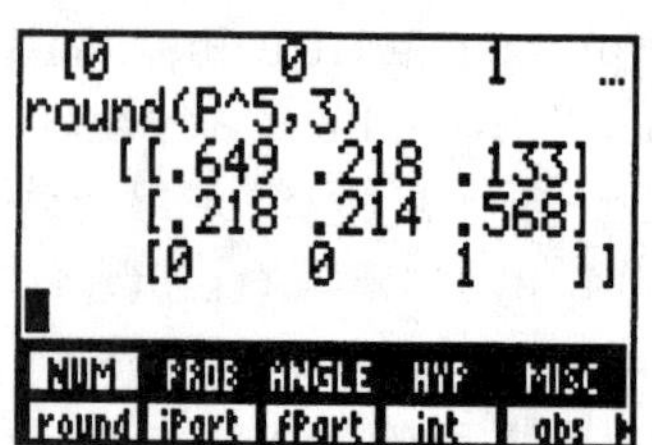

[MATH] [F1]). Once you are in this menu, press [F1] (round) [ALPHA] [P] [∧] 5 [,] 3 [)] [ENTER]. This command rounds only the necessary values to 3 decimal places.

Now press [2nd] [ENTER] to bring back the last issued command, arrow to the left so that the cursor is over the 5, and type 25 [,] 3 [)] [ENTER] to compute P^{25}. Repeat this process and compute P^{100} and P^{150}.

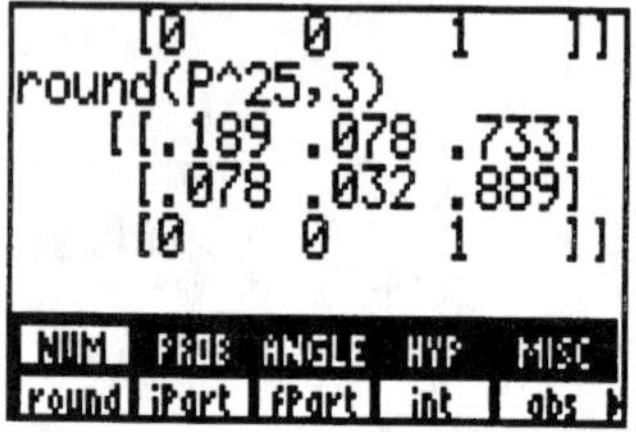 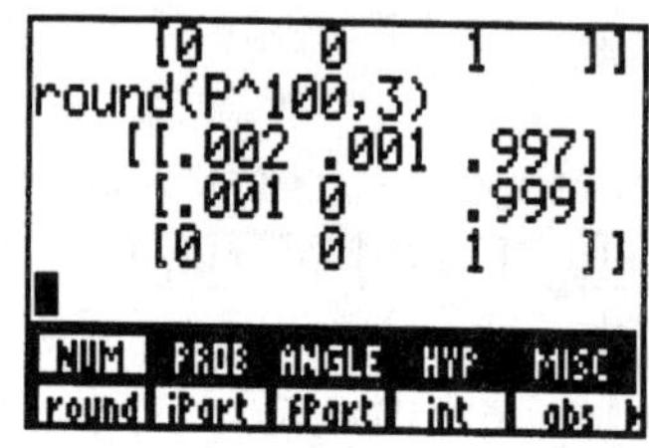 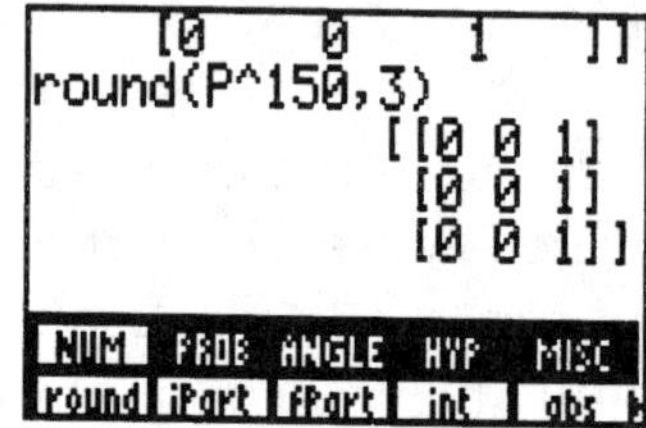

The 1's at the end of the first and second row of P^{150} indicate that after 150 years, there is a 100% chance that those S&L's that are currently profitable and those operating at a loss, respectively (i.e., *all* S&L's), will be bankrupt!

Now Compute P^{155}. What does this matrix tell us? If all S&L's are bankrupt in 150 years, they would still be bankrupt in 155 years (*why?*). This is why the matrices P^{155} and P^{150} are the same. Can you find the number of the earliest year in which this phenomenon would first occur?

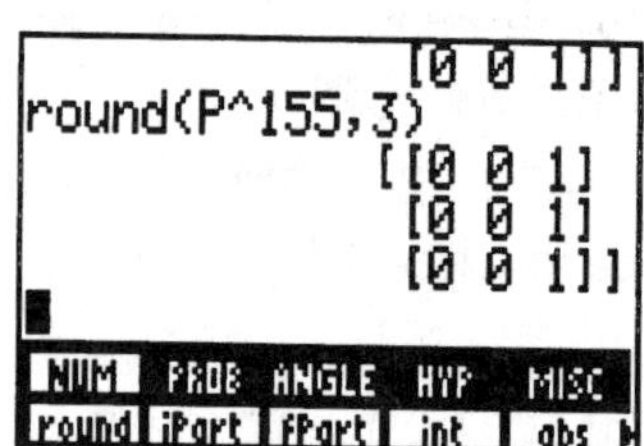

9.3 Long Range Behavior of Regular Markov Systems

This section is basically an application of matrices and systems of equations, as finding a steady-state vector theoretically amounts to nothing more than solving an associated system of equations. After setting up the corresponding system of linear equations, you can find a matrix solution to the system using matrix row operations on the TI-86 or by using the reduced row echelon form feature `rref` on the calculator.

In Example 2 *Calculating the Steady-State Vector* in the text, the transition matrix is given by $P = \begin{bmatrix} 0.8 & 0.2 \\ 0.1 & 0.9 \end{bmatrix}$ and corresponding system is given by

$$\begin{array}{rl} x + y &= 1 \\ -0.2x + 0.1y &= 0 \\ 0.2x - 0.1y &= 0 \end{array}$$

To find the steady-state vector $v_\infty = [x \quad y]$ theoretically, input the augmented matrix $\begin{bmatrix} 1 & 1 & 1 \\ -0.2 & 0.1 & 0 \\ 0.2 & -0.1 & 0 \end{bmatrix}$ of the system into your TI-86 and name it A. Verify on the home screen that the matrix entries have been inputted correctly.

Press [2nd] [MATRX] [F4] (OPS) [F5] (rref) [ALPHA] [A] [CUSTOM] [F2] (▶Frac) and then [ENTER] to obtain the reduced row echelon form of matrix A with fractional entries. The result shows that $x = 1/3$ and $y = 2/3$, and hence the stead-state vector and long-term transition matrix is $v_\infty = \begin{bmatrix} \dfrac{1}{3} & \dfrac{2}{3} \end{bmatrix}$ and $P^\infty = \begin{bmatrix} \dfrac{1}{3} & \dfrac{2}{3} \\ \dfrac{1}{3} & \dfrac{2}{3} \end{bmatrix}$, respectively.

```
                [-.2 .1  0]
                [.2  -.1 0]]
rref A▶Frac
               [[1 0 1/3]
                [0 1 2/3]
                [0 0  0 ]]
 abs  ▶Frac   xⁱ
```

Determining if a Markov System is Regular

To determine if a Markov system with transition matrix P is regular, you must determine if there is some power of P that contains no entries that are zero. The TI-86 can be used to aid us in this determination. In Example 4 from the text, *Regular Markov System*, part (a) gives the transition matrix

$$P = \begin{bmatrix} 0.5 & 0 & 0.5 \\ 0.6 & 0 & 0.4 \\ 0 & 1 & 0 \end{bmatrix}.$$ Input this matrix into the calculator as P and verify on

the home screen that P has been defined correctly. Make any changes to P if necessary.

After verification, the matrix P becomes your last answer. Press [2nd] [MATH] [F1] [F1] [2nd] [ANS] [x^2] [,] 3 [)] to find the square of P, rounding entries to three decimal places.

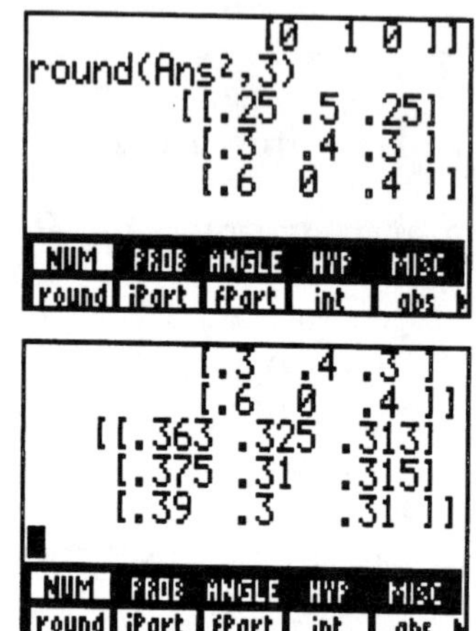

After noticing that P^2, the matrix just found, has a zero entry, continue to find P^4 by pressing the [ENTER] key again (remember, this tells the calculator to perform the last issued command again, using the most recent answer, P^2, as ANS.

Thus, you have squared P^2 and therefore obtained the matrix P^4. Since this matrix has no zero entries, the system is regular.

For the system in part (b), with transition matrix $P = \begin{bmatrix} 0.6 & 0.4 \\ 0 & 1 \end{bmatrix}$, input this

matrix into the TI-86's matrix editor and verify the correctness of the entries on the home screen.

Perform the same analysis as just done for the transition matrix of part (a). Recognizing that there seems to a zero entry in each subsequent matrix P^2, P^4, etc., you can conclude that the system is not a regular system. Continue to press [ENTER] to compute higher even powers of P to convince yourself of this.

9.4 Absorbing Markov Systems

Although there are some extremely interesting problems in this section, there are no new calculator skills presented in this section.

You're the Expert – Predicting the Price of Gold

There are no new calculator skills in this section.